# The Geographical Association

## The First Hundred Years
## 1893–1993

Patron of the Centenary Celebrations, HRH The Princess Royal

# W. G. V. Balchin, MA. PhD.

Emeritus Professor of Geography
University College of Swansea
Fellow of King's College London

The Geographical Association
343 Fulwood Road Sheffield S10 3BP
Tel: 0742 670666

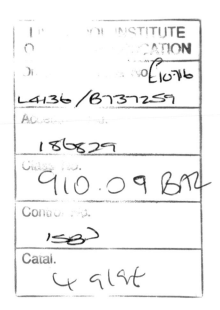
Front cover: Geography lessons in the 1990s. Fieldwork is of great importance.
Photo: John Bebbington, FRPS, Field Studies Council.

Back cover: Geography lessons at the beginning of the century. Top: Hague Green School, Bethnal Green, 1908. Bottom: Rushmore Road School, 1908. Photos: Greater London Photograph Library.

The views expressed in this publication are those of the author and do not necessarily represent those of the Geographical Association.

A grant in aid of the publication of this Centenary History from BP Exploration is gratefully acknowledged.

ISBN 0 948512 57 1

# Contents

After an unparalleled research we are now able to reveal the great Secret of Geography. The secret is that without Geography you would be quite lost: you wouldn't know where you were, or whether you were a native or British, or where the nearest mangrove swamp was, or anything . . .

(Geography Part I in W. C. Sellar and R. J. Yeatman *And Now All This*, 1932)

# Preface

Although by no means the oldest member of the Geographical Association I joined in 1937 and became actively engaged in its organisation in 1951 when appointed to the post of Conference Organiser. For many years tradition had decreed that the Hon. Conference Organiser had to be a member of the Joint KCL/LSE staff as the Annual Conference was always held at the London School of Economics: when Professor Beaver left LSE to take up the new Chair of Geography at the University College of North Staffordshire (later Keele University) the post of Conference Organiser crossed the Strand to King's College and fell to me until 1955 when my resignation had to follow having been appointed to the Foundation Chair of Geography in Swansea in 1954.

What then looked like the end of a story proved in fact to be the beginning of a long, continuous connection with the Association. The mid-fifties were marked by a succession of unexpected deaths amongst the elderly Trustees of the time with the result that Dr. Alice Garnett, in some desperation, sought the youngest professor of the period as one of the replacements. Trustee posts were then life appointments and involved permanent service on the Council and Executive. This appointment continued until the revised Constitution of the 1970s limited the tenure of all posts and retirement in my case was imposed in 1977. But I was retained on Council until 1981 owing to the special relationship which had existed with the Ordnance Survey. This terminated when the OS Consultative arrangements changed as a result of the Serpell Committee Report.

Experience of the working of the Geographical Association at the grass roots level has also been obtained as a result of an immediate installation as President of the Swansea Branch on arrival there in 1954: this office was to last until retirement in 1978. Presidency of the whole Association in 1971 also provided additional insight into the work of the Association. By the early 1980s I thought I had at long last escaped but in 1990 came a call to become President of the Bradford Branch and also to write the history of the Association! As a result of the long, close contact with

the formal working of the Association I have had the good fortune to know personally many of our now legendary founders and to appreciate at first hand the magnificent work they undertook to establish geography so firmly in our schools and universities. It was therefore with pleasure that I received the request to write a concise history of the Association for the centenary celebrations in 1993.

Here was a wonderful opportunity to place on record our collective thanks for all the contributions and achievements of the founders to whom we owe so much. Sadly both the professional and public memory is short and only too often the work of the pioneers is forgotten in the admiration bestowed on the work of their successors. We also need to record the great debt which the Association owes to its countless contributors who have given their time, energy and services, in writing and preparing texts, lectures and pages for publication without any question of fees or royalties. The Association is indeed a great voluntary organisation in which members have worked freely for the good of their colleagues and the subject in general. Happily it continues to survive in this form despite the commercialisation of so many educational activities. Long may it remain so.

I am particularly indebted for assistance received in the compilation of the history to the Headquarters staff of the Association in Sheffield, especially Miss Julia Legg and Mrs. Noreen Pleavin, also to Mrs. C. Kelly, lately archivist at the Royal Geographical Society in London, and to Mr. G. B. Lewis of the Department of Geography in the University College of Swansea who assisted with the maps and diagrams. Additionally my thanks are due to Professor Norman Pye, Patrick Bailey, James Hindson and Geoffrey Sherlock for many improvements in the text and presentation. Very special thanks are also due to the BP Exploration Operating Company Ltd. for their interest in the project and for a substantial grant in aid of publication.

*W. G. V. Balchin*
*January 1993*

# List of Illustrations

Photographs
Frontispiece: Christ Church, Oxford, Common Room

Figures

Frontispiece: The Common Room at Christ Church Oxford in which the meeting of 20th May 1893 was held and the Association founded.

# The First Phase

## The Founding

The Geographical Association has been primarily interested in the development of the teaching of geography and it has therefore been concerned both with the evolution of geographical studies and in changes in attitudes to education during its hundred years of existence. Its emergence at the end of the nineteenth century relates to both these factors, operating through the activities of certain remarkable and memorable individuals.

The background to what occurred in the 1890s lies in the fact that geography as we know it today did not exist in the schools or universities of Great Britain in the nineteenth century. Geography was of course in evidence at both Oxford and Cambridge during the medieval period but more often than not it was in the guise of 'cosmography', 'astronomical geography', 'mathematical geography' or surveying, with the geographer taking on the role of map-maker (an activity which became cartography in the 1830s). Oxford can claim such well known names as Hakluyt, Camden, Carew, Abbot, Prideaux, Carpenter, Pemble, Burton and Halley (Baker, 1963); and Cambridge, Cunningham and Dee (Taylor, 1930).

Towards the middle of the eighteenth century however a decline of interest in geography occurs in the established universities and this continues throughout the nineteenth century. The revival towards the end of the century was due entirely to the initiative taken by a small group of enthusiasts operating within the Royal Geographical Society.

Although the subject was in such a parlous condition in both Oxford and Cambridge as well as in the schools, public interest was not extinguished. Much of the century was concerned with efforts to solve the great exploration mysteries such as the North West Passage, the North East Passage, the sources of the Nile, the North and South Poles, and the opening up of Africa and Australia. The Royal Geographical Society therefore flourished and there was a great demand for books on travel and exploration. For the academically minded, books such as Mary Somerville's *Physical Geography* (1848), which reached six editions by 1870, were available. Human geography was then known as political geography and this found a place in many encyclopedias. The 8th Edition of the *Encyclopedia Britannica* (1853) had many geographical articles, whilst E. Reclus's 15 volume *Geographie Universelle* (1878–1894) supplied systematic accounts of most countries. T. H. Huxley's *Physiography* (1880) continued the Mary Somerville tradition: other well known authors of the period included A. C. Ramsay (*Physical Geology and Geography of Great Britain*, 1863) and A. Mackay (*Manual of Modern Geography*, 1870). The University of London had also shown an interest in the subject as there were Professors of Geography at University College from 1835–37 (A. Maconochie) and Kings College from 1863–76 (William Hughes). But at Oxford and Cambridge and in the public and grammar schools geography was completely subservient to classics, whilst in the elementary schools there was little room for anything but the three Rs. In the few schools where geography was taught it rarely went beyond the 'capes and bays' approach and there were few if any satisfactory school textbooks. It is said that the geologist Professor Geikie asked Dr. H. R. Mill, Librarian at the Royal Geographical Society, to write an elementary geography and sent him 25 textbooks as examples of how *not* to do it! J. R. Green's *Short Geography of the British Isles* (1879) is perhaps one of the few exceptions of this period.

It is against this background that several influential and remarkable members of Council of the Royal Geographical Society reacted. The first to move was Sir Francis Galton (Secretary of the Society from 1856–63 and intermittent periods between 1871–92) who persuaded the Council to offer prize medals for

1

Photo 1. Douglas William Freshfield, 1845–1934. Explorer, mountaineer and educationalist. Founder member and President of the Geographical Association 1897–1911.
Photo: Royal Geographical Society.

At the same time Douglas Freshfield resurrected the submissions which the Royal Geographical Society had made at the time of the Royal Commissions into University affairs in the 1850s and 1870s, and advanced them a stage further with the proposal that the Society should help fund the cost of appointments in geography at Oxford and Cambridge.

Another critical name now appears quite independently on the scene in the shape of Halford John Mackinder who arrived at Oxford in 1880 with a Junior Studentship at Christ Church. He began his university studies with Natural Sciences obtaining a First in 1883. This was followed by a year studying Modern History and Geology in order to compete for the Burdett Coutts Scholarship, which he won. After this he read Law, being called to the Bar at the Inner Temple in 1886. During the latter period he was lecturing for the Oxford University Extension movement on what he called 'the new geography' which was based on his background studies of physiography, climatology, hydrology, biology and history. Mackinder's personality was to prove a vital factor in the development of both geography and the Geographical Association. 'His

competition amongst the public schools. The medals unfortunately consistently went to the same schools and the scheme failed to improve the standard of teaching: it was eventually dropped in 1884. Meanwhile Francis Galton had been joined by James Bryce, Douglas Freshfield (Secretary of the Society from 1881–93) and 'Bates of the Amazon' (Assistant Secretary).

This group, led by Douglas Freshfield, persuaded the Council to institute an enquiry into the teaching of geography both in England and overseas. The task was given to F. Scott Keltie in 1884 and his report, which was to trigger the subsequent development of geography in Great Britain, became available in 1885. The report revealed the great advances which had been made overseas in the teaching of the subject— at least 45 chairs of geography were noted in continental universities—and also listed an impressive collection of books, maps, atlases and apparatus available in support of the teaching.

Photo 2. Sir John Scott Keltie, 1840–1927. President of the Geographical Association 1914.
Photo: Royal Geographical Society.

handsome presence, with flashing eye and a gift of oratory that has rarely been bettered, made him an inspiring lecturer who drew large undergraduate audiences' (H. J. Fleure).

News of this brilliant and stimulating lecturer and the 'New Geography' reached Francis Galton at the Royal Geographical Society and Mackinder was invited to give his famous 'Scope and Methods of Geography' lecture to the Society in 1887: the Society immediately realised that they had the right man for the proposed Oxford Readership. Agreement was rapidly reached with Oxford and Mackinder was installed as a Reader for a five year tenure. An appointment was made in the following year for Cambridge but the initial choice was less successful than at Oxford. The battle for geography at Oxford and Cambridge was not yet won however for the exponents had next to build for themselves and the subject an audience, a reputation, and then justify an examination. We owe a great debt to Mackinder for achieving this state of affairs fairly rapidly at Oxford: and later to J. Y. Buchanan and H. Yule Oldham for establishing the subject successfully at Cambridge (Balchin, 1988).

Meanwhile a few progressive teachers of geography in the public schools were endeavouring to keep pace with the new ideas and methods of teaching being proposed by Mackinder and Scott Keltie, one of which was the use of the lantern projector. It is difficult now to appreciate the heated discussions which the use of the lantern projector generated in the 1890s. Douglas Freshfield was also an enthusiastic exponent of the lantern but found much opposition to its use at the Royal Geographical Society by certain Council members who objected to these 'Sunday school methods'.

In the public schools B. Bentham Dickinson of Rugby was a leading lantern advocate who wanted more lantern slides than he could produce himself: as a result he had the bright idea in 1893 of attempting an exchange scheme with other teachers. The idea was sent to the Royal Geographical Society who passed it to Mackinder at Oxford. It is from this correspondence that we trace the origins of the Geographical Association, for on April 4th 1893 a letter was circulated to a number of

Photo 3. Sir Halford J. Mackinder, 1861–1947, Founder member; Chairman of Committee 1908–1913; Chairman of Council 1913–1947; Trustee 1914–1917 and 1919–1947; President 1916.

public schoolmasters inviting them to a meeting at Christ Church Oxford to discuss the proposal. The letter was signed by Douglas W. Freshfield, Honorary Secretary for the Royal Geographical Society; by T. Field, the Headmaster of King's School in Canterbury, who had been advocating the use of the lantern in the teaching of history; by H. J. Mackinder, as Reader in Geography at Oxford; by B. B. Dickinson, the Geography Master from Rugby School; and C. B. Hewitt, an Assistant Master at Marlborough College. Eleven individuals were present at the meeting on May 20th 1893—Mackinder and ten public schoolmasters. Twelve other schoolmasters sent apologies but expressed interest. The discussion on the use of the lantern eventually broadened into the larger question of the formation of an association to promote the teaching of geography. The lead here was taken by H. J. Mackinder who made a definite proposal from the chair which was

*Oxford, 4th of April, 1893.*

Dear Sir,

In connection with schemes for the improved illustration of Geographical and other teaching in Schools, which have lately been circulated by Mr. Field of Canterbury and also by Mr. Dickinson of Rugby, it has been arranged to hold a Meeting of School-masters and others interested at Oxford, on Saturday, the 20th of May, at 4.30 p.m. in the New Common Room, Christ Church.

It is hoped that the discussion will lead to the adoption of a practical scheme.

We trust that you may find it possible to be present, and to place your experience at our disposal.

We remain,

Yours faithfully,

DOUGLAS W. FRESHFIELD,
(Hon. Sec. to the Royal Geographical Society.)

T. FIELD,
(Head-Master of King's School, Canterbury.)

H. J. MACKINDER,
(Reader in Geography in the University of Oxford.)

B. B. DICKINSON,
(Assistant-Master in Rugby School.)

C. E. B. HEWITT,
(Assistant-Master in Marlborough College.)

Kindly reply to
B. B. Dickinson, Bloxam House, Rugby.

Figure 1. A facsimile of the Notice which called the Oxford meeting of 20th May 1893.
Geographical Association archives.

Photo 4. The staff of the Oxford School of Geography with the first candidates for the Diploma in Geography, outside the Old Ashmolean, June 1901.
Back row (from left to right): Dr. H. N. Dickson, Lecturer in Physical Geography, E. C. Spicer, W. Stanford, W. Bisiker, Dr. A. J. Herbertson, Assistant to the Reader and Lecturer in Regional Geography.
Front row: Miss J. B. Reynolds, H. J. Mackinder, Reader in Geography.
Photo: Royal Geographical Society and Oxford School of Geography

unanimously adopted. Dickinson, somewhat to his dismay, found himself appointed honorary secretary of the new Association. The Geographical Association thus emerged largely as a result of the pioneer work and key interests of Francis Galton, Douglas Freshfield, J. Scott Keltie, B. B. Dickinson and H. J. Mackinder.[1]

[1]Concise biographies of the main individuals contributing to the early history of the Geographical Association will be found in appendix K.

# The Second Phase

## B. B. Dickinson as Honorary Secretary
## 1893–1900

Photo 5. B. Bentham Dickinson, Died 1941. Founder member; Honorary Secretary 1893–1900; President 1930.

Photo: Royal Geographical Society.

Acting on the meeting of 20th May 1893 invitations were despatched to potential members of the newly formed association: a yearly subscription of five shillings was quoted. On Thursday 3rd August 1893 a further meeting was held to elect a managing committee and officers. Only six members attended this meeting but they knew that they had 35 subscriptions so they proceeded to elect a committee consisting of Dr. Hugh Robert Mill (Librarian at the Royal Geographical Society) as

Chairman; B. B. Dickinson (of Rugby School) as Secretary; C. E. B. Hewitt (of Marlborough College) as Treasurer; supported by J. Robinson (of Dulwich College) and the Rev J. Ll. Dove (of Haileybury College). The committee was able to note that the Royal Geographical Society (through Dr. Mill) would allow Geographical Association members to receive the newly established *Geographical Journal* at a reduced cost, whilst Messrs. George Philip and Son of Fleet Street proposed a discount on lantern slides. The Royal Colonial Institute also joined as a Corporate member with a three guinea subscription (a princely sum in 1893:). Mr. Dickinson had also prepared a catalogue of 238 specially made slides. The Geographical Association was beginning to become active.

The first general meeting at the Royal Colonial Institute followed on 21st December 1894 when the reported membership was 50. Three further names were added to the committee—J. S. Masterman, G. C. Harrison and E. R. Wethey, of whom Masterman was to prove the most significant. All these moves were favourably noted by the Royal Geographical Society with whom contact was maintained at all times. The Society did not wish to become involved in the intricacies of teaching at school level as it had as much as it could cope with at the university level. Subsequent to the promotions at Oxford and Cambridge, Royal Geographical Society assistance was also provided at the Universities of Manchester, Edinburgh and Wales. The link with the Geographical Association was arranged first by means of Dr. H. R. Mill, the Society Librarian, who remained on the Association Committee from 1893 to 1901, and secondly by Douglas Freshfield, who had worked tirelessly for improvements in the teaching of geography ever since becoming Secretary of the Royal Geographical Society in

1881. Freshfield was in many ways ahead of his time supporting not only lantern projection but also the admission of ladies to the Fellowship of the Society. The historic argument about the admission of ladies in the early 1890s eventually led to Freshfield's resignation from the Secretaryship and the production of his memorable couplet:

The question our dissentients bellow
Is 'Can a lady be a fellow',
That, Sirs, will be no question when
Our fellows are all gentlemen.

The Society's loss of Freshfield was however the Geographical Association's gain as he was able to devote more time to the new body, becoming its first President in 1897 and remaining in this office until 1911.

The transition from a small group of teachers exchanging lantern slides and meeting occasionally, to a fully active Association, took place after criticism in the *Educational Review* of January 1895 where it was pointed out that the work of the Geographical Association was of interest to over 700 grammar schools, most of which were not aware of its existence. Much of the implied need for expansion was outside Dickinson's secretarial ability as a working teacher but a solution was found in the appointment of J. S. Masterman as an assistant after his retirement in 1894 from his teaching post at University College school. This arrangement continued until the changes of 1900 took place and Masterman became Treasurer.

During 1894–5 widespread enquiries were made amongst geography teachers on four major points.
1. Should examination papers in geography be prepared and/or reviewed by experts?
2. Should a knowledge of physical geography be an essential feature in a course in geography and in any subsequent examination, and if so what should be the syllabus?
3. Should one ask for a knowledge of the whole world in general or for a more detailed knowledge of a continent or region?
4. Should geography be a compulsory subject for some competitive examinations?
Needless to say the replies ranged widely but the general consensus was much as might now be expected: examination papers should be set by experts; physical geography was an essential part of the teaching; general world geography with special study of selected regions was the ideal to aim for; and geography should be compulsory in certain examinations.

Support for these conclusions came from the Royal Geographical Society, the Royal Colonial Institute, and the Teachers Guild (with whom the Association became closely connected and received much support from 1895 onwards). Further suggestions were submitted by the Royal Scottish Geographical Society, The Association of Teachers in Secondary Schools in Scotland, and the Manchester Geographical Society. The end result was a Memorandum which went to all examining authorities urging that physical geography should be the basis of all geographical teaching, that general world geography should be supplemented by specialist study of selected regions, and that geography and history should have separate papers with equal marks.

The reactions of the examining authorities was varied. The Victoria University of Manchester and the Oxford and Cambridge Schools Examination Board responded by adjustments to their regulations. The teachers College of Preceptors claimed to be meeting the Association's proposals. The Scottish Education Committee did not agree. London University did nothing. The Civil Service Commission produced the argument that to emphasise physical geography would give unfair advantage to those also offering geology—the Commission having failed to realise that the geographers were concerned with relations between physical and human aspects rather than physical phenomena *per se*.

Another problem which had exercised the minds of the Committee for some time was the question of a suitable syllabus for school use. The situation was so fluid at the end of the nineteenth century however that the Committee wisely decided not to lend their authority to any syllabus, preferring to encourage individual teachers to produce their own ideas or method derived from practical experience. The Association however published a number of specimen schemes and gave publicity to the discussions on this topic which appeared in the American *Journal of School Geography*.

Supported by the Preparatory schools the

Geographical Association next asked Head-masters of the Public Boarding Schools to make geography a subject in their entrance examinations, but the request was declined. The success of the lending scheme for lantern slides next led to proposals, along with the teachers' Guild, for a lending library of maps and pictor-ial illustrations, but this failed for practical rea-sons. The idea was to be resurrected forty years later but once again failed to mature. One very successful venture was Dr. Mill's invaluable compilation *Hints to Teachers of Geography on the Choice of Books for Research and Reading* (1897). As Librarian of the Royal Geographical Society Dr. Mill was ideally placed to prepare such a publication.

As originally conceived the Geographical Association was open to membership from Scotland as well as England and Wales but recruitment north of the border was slow (it is stated in several places that the five shilling annual subscription was a great deterrent!).

Photo 6. Dr. Hugh Robert Mill, 1861–1950. Member of Committee 1894–1901; President 1932. Photo: Royal Geographical Society

Although a Scottish Branch emerged in 1898 and submitted proposals for improvements in examinations in geography in Scotland its activities appear to cease after February 1899. It is perhaps significant that A. J. Herbertson, then in Edinburgh, was associated with these activities before his move to Oxford. Subse-quently Scottish representation on the main committee was strengthened. An independent Scottish Association of Geography Teachers did not emerge until the 1970s.

Additionally there were negotiations with the publishers of school atlases in an effort to improve some of the cartographic representa-tions, so that as the end of the century approached the Geographical Association had become an active force in British Geography. Much doubtless relates to the creation of a Presidency for the Association in 1897 and the appointment of Douglas Freshfield to this post. His interest in promoting the subject had continued unabated, and we see another of his forward looking views operating when in 1900 the Geographical Association opened its mem-bership to all teachers of geography irrespective of age, sex or type of school. The Committee was also strengthened by the addition of pow-erful supporters such as Sir Richard Gregory, Sir Frederick Ogilvie, G. G. Chisholm, R. D. Roberts and A. W. Andrews (a University Extension lecturer in Oxford and London).

Although we now think primarily of lantern slides (Fig. 2) in relation to the early work of the Association it is worth recalling that there were other visual-spatial aspects of teaching that also received attention. As well as the ordinary photographic lantern slides teachers had realised the power of the map slide. Also, although globes were probably more widely available then than nowadays, wall maps were notoriously poor, whilst colour differentiation for orography was then in its early stages and giving rise to a good deal of discussion. Contour lines were still relatively new and their full value had yet to be realised.

Additionally members of the Geographical Association could look forward to the special lectures at the annual meetings, the most notable being those given by H. J. Mackinder on 'Geography as Training of the Mind' and those by B. B. Dickinson on 'The Use of the Lantern Slide'.

Figure 2. Two slides from the early archives of the Geographical Association used to promote the use of lantern slides.

# The Third Phase

## A. J. Herbertson as Honorary Secretary
## 1900–1915

The turn of the century witnessed considerable changes in the disposition of the leading geographers in the country and these changes had an impact on the development of the Geographical Association. In 1899 the growing popularity of geography in Oxford led to the creation of a 'school of geography' with H. J. Mackinder appointed as Head and continuing as Reader: additionally A. J. Herbertson was appointed an 'assistant', with H. N. Dickson and G. B. Grundy as lecturers in physical and ancient geography. Herbertson went to Oxford via a lectureship in geography at Manchester where the Royal Geographical Society had also been supporting the subject: he had become known for the help given to Bartholomew in the production of the Atlas of Meteorology.

In his new position Herbertson began to take an increasing interest in the work of the Geographical Association. Dickinson by this time was beginning to feel the strain of the Secretary's post despite the help given by Masterman. The 'Diagram Company' (producing lantern slides) run by Andrews and Dickinson took up an increasing amount of time. As a result in 1900 Herbertson succeeded Dickinson as Secretary, Masterman became Treasurer, but Freshfield remained President of the Association. Another change came in the following year when Dr. H. R. Mill was appointed Head of the British Rainfall Organisation and felt that he would have to resign from his commitments both in the Geographical Association and the Royal Geographical Society. He was made an Honorary Member of the Geographical Association and became its President many years later in 1932, maintaining his interest in the Association until his death in April 1950 at the age of 89 years.

The impact of Herbertson as Secretary under the Presidency of Freshfield was imme-

Photo 7. Professor Andrew John Herbertson, 1865–1915. Honorary Secretary 1900–1915.
Photo: Geographical Association Archives.

diate and far reaching. The Association membership was opened to all geography teachers in every type of institution and plans were prepared to publish a journal to keep members in touch with developments in the teaching of geography. The journal was a financial risk but a guarantee provided by Mr T. G. Rooper, an H.M. Inspector of Schools, ensured publication for at least three years by which time it was thought that the journal would have become established. Happily this optimism was to prove correct and the first issue of *The Geographical Teacher* under the joint editorship of W. A. Andrews and A. J. Herbertson appeared in October 1901. The editorial partnership was to continue for the next three journals but in 1903 A. J. Herbertson became the sole Editor until his early and unexpected death in 1915.

The Journal of the Association enshrined the views and thoughts of the pioneers of the changed approach to the teaching of geography and helps us to chart the gradual development of the subject throughout the rest of the century. The first number of *The Geographical Teacher* (free to members, price one shilling to non members) contains an enthusiastic introduction by the President, Douglas Freshfield, and articles on Methods of Teaching Geography by T. G. Rooper, The Use of Maps by A. W. Andrews, Teaching the Geography of the World by A. J. Herbertson, Photography as an Aid in Teaching Geography by C. C. Carter, and contributions on examination papers, geographical literature and recent maps and apparatus. These articles and other contributions determined the character of the journal for many issues to come.

It is significant that references to field work occur in no less than six of the articles in this first number thus foreshadowing the remarkable rise of this hallmark of British geographical teaching. It is about this time that Cowham's formal school field excursions from Croydon to Godstone and Reynolds' excursions from the Friend's High School of Kendal to Grassmere were becoming widely known. Joan Reynolds (a cousin of the Kendal Reynolds) notes in her *Geographical Teacher* article:

> while these excursions are still innovations we must remember how important it is that such as are carried out should be successful from all points of view. They must prove themselves to have a direct as well as an indirect educational value; or British parents, who will have to bear the extra expense and are slow perhaps to perceive the value of this new form of training, will taboo them as a needless extravagance . . .
>
> Attention to all these matters involves, I know, a great deal of patient thought and trouble, and yet I feel it is well worth it and thereby we may realise in time the ideal held up by Ruskin when 'The country will become an outer, and uncovered class-room, a Divine Museum utilised by our teachers'.

A. J. Herbertson as co-editor and after 1903 as sole editor played an immense part in the development of the subject. Not only did he control the content of *The Geographical Teacher* but he also had the means of propagating his own brilliant ideas; there is little doubt that his voluntary work for the Association contributed greatly to the rise in eminence of the Oxford School of Geography at the beginning of the century. The membership of the Association also began to rise steadily as the effects of the 1902 Education Act, which created the County Secondary School structure in England, were felt—at long last geography specialists were in demand by the State system. The annual meeting of the Geographical Association at the College of Preceptors in London on 15th January 1902 records a membership of 202.

The pattern of activity for the Association in the next few years settled down into the publication of *The Geographical Teacher* three times a year, an Annual Conference in London in January, and of course the loan scheme for slides. The Annual Conference migrated around London from the College of Preceptors to the South West Polytechnic, the Royal Colonial Institute and the Medical Examination Halls, arriving at the London School of Economics in 1907. This venue probably reflected other changes that had taken place in the university world of geography. H. J. Mackinder had resigned his Oxford Readership in 1904, having become Director of the London School of Economics in 1903 (where he had been a Lecturer in Geography since 1895, during which time he was also Principal of the newly founded University College of Reading as well as being Reader in Geography at Oxford. He is reputed to have kept dinner suits in Oxford, Reading and London to save trouble!) Herbertson succeeded Mackinder at Oxford and this along with further developments in the Association necessitated more changes.

By 1905 the Geographical Association membership had reached 486 (of whom 153 were ladies) and a new development occurs with the appearance of branches in South London (March 17 1904), Bedfordshire (May 13 1905), and Bournemouth (May 17 1905). The need for assistance for Herbertson became urgent

Photo 8. Professor John Frederick Unstead, 1876–1965. Honorary Librarian 1908–1914; Vice President 1922–26 and 1962–65.
Photo: Birckbeck College.

and J. F. Unstead, recently made a Lecturer in Geography at Goldsmiths College, was appointed as an Honorary Corresponding Secretary. Significantly Herbertson retained control of *The Geographical Teacher* which by 1905 had become well established, appearing regularly three times a year in the Spring, Summer and Autumn.

As might be expected the first volumes of *The Geographical Teacher* contain a predominance of articles by the early stalwarts of the Association, A. W. Andrews, A. J. Herbertson, C. C. Carter, B. B. Dickinson, H. J. Mackinder and J. F. Unstead figure prominently discussing problems such as the relative place of local studies and world studies, the use of maps and globes in geography, the regional approach to geography, and the use of the lantern. Contributions were also sought and

published from eminent authorities in cognate subjects such as the Rt. Hon. J. Bryce MP (Geography in Education), Sir Archibald Geikie FRS (The Use of Ordnance Maps), the Hon. Sir John Cockburn (The Australian Commonwealth) and Professor Salisbury (Geography in the University of Chicago).

One very useful item for the research worker studying the development of geography in the British Isles is the inclusion in the early volumes of details of the courses, syllabuses, examinations, staff, and teaching arrangements at universities where geography existed. Eventually with the growth of the subject this feature had to be dropped as it began to occupy too much space. The beginnings and the growth are however clearly recorded. Additionally notes on new geography books, atlases and maps rapidly gave way in 1903 to more detailed and informative reviews. Detailed critiques of school examination papers are also found in the early volumes and these appear to have had a considerable influence in vetting subsequent examination papers.

*The Geographical Teacher* also provides an insight into the difficulties that confronted geography teachers at the beginning of the century. An article by A. T. Simmons on 'Geographical Laboratories' in 1908 contains the following telling paragraphs: on page 261 dealing with the equipment for measuring distance we read:

> The material equipment for work of this kind need not be elaborate. All we require is a watch: and despite sadly inadequate salaries most teachers possess a watch: and even if they do not, since we can purchase one for five shillings, we shall not be making an excessive demand on the school exchequer if we ask for one as a piece of geographical apparatus. Most boys' schools possess a tape measure of the kind required, which they use for the purpose of games: and the same is true of most girls' schools . . .

and on page 266 dealing with globes we read:

> And on this I would lay special emphasis, a small globe for each pupil. You can purchase for eighteen pence really very passable globes, about 4 inches in diameter, suitably supported at the right inclination. If this is too expensive

you can obtain for one penny, at most confectioners, a metal terrestrial globe divided into hemispheres and filled with chocolate.

(Pennies were of course pre-decimal with 240 to the pound!)

In the autumn of 1905 *The Geographical Teacher* contained the famous paper by A. J. Herbertson on the World's Major Natural Regions. With various revisions this paper was to become a valuable teaching tool and was to influence geography teaching profoundly for the next forty years: it represents an effort to achieve a synthesis of physiographic, climatological and pedological factors, as expressed through vegetation, for the whole world. It clearly reflects Herbertson's work on the Bartholomew Atlas of Meteorology during the 1890s and was first presented to the Research Committee of the Royal Geographical Society in February 1904 and published in the *Geographical Journal* in 1905. In terms of its impact upon the teaching of the subject it must surely rank alongside that of Mackinder's 'Scope and Methods of Geography' paper of 1887. Needless to say Herbertson's new ideas sparked opposition, notably by L. W. Lyde, the newly appointed (1903) Professor of Economic Geography at University College London who argued for the division of the continents into political units rather than natural regions which he thought too subjective and indefinite. Now an almost forgotten name, Professor Lyde dominated the textbook market at the beginning of the twentieth century. His *Man on Earth* and *Men and His Markets*, actually written in the 1890s pioneered the concept of geography as primarily the study of man in his environment. The present writer's introduction to geography was by way of Lyde's *The Continent of Europe*, first published in 1913 and subsequently followed by many up-dated editions.

In the same year, 1905, another important event took place when H. R. Mill gave the copyright of his *Hints to Teaching Geography* to the Geographical Association on condition that a new edition be prepared. The result was the *Guide to Geographical Books and Appliances* which with the help of J. F. Unstead, appeared in 1909: the Geographical

Photo 9. Professor Lionel William Lyde, 1863–1947. Trustee 1914–1947.
Photo: Royal Geographical Society.

Association thus began its own book publication record with members writing without royalties for the benefit of other members. This policy was to have a profound effect in the Association's subsequent history.

The last years of Douglas Freshfield's Presidency were marked by further changes in the Geographical Association's personnel, partly as a result of his explorations abroad. G. G. Chisholm, then a Lecturer in Geography at Birkbeck College London became Chairman of Committee in 1907 in order to deputise for Freshfield but was almost immediately forced to relinquish the office on his translation to Edinburgh in 1908. The Association thereupon elected H. J. Mackinder as Chairman of Committee and, despite all his subsequent activities, Mackinder held this position until his death in 1947 thus establishing a truly remarkable involvement with the Association over half a century.

The operational headquarters of the Geographical Association continued in Oxford where in 1910 A. J. Herbertson was elevated to a personal chair: already Honorary Secretary of the Association and Editor of its Journal

Photo 10. James Fairgrieve, 1870–1953. Member of Committee 1909–1913; Member of Council 1913–1953; Chairman of Council 1947–48; President 1935.
Photo: Geographical Association Archives.

this further advancement placed him in a unique position to influence the development of the subject and the Oxford School of Geography rapidly became pre-eminent in British Geography. We also note in 1909 the first appearance in the Committee list of the name of James Fairgrieve who was later to play such an important part in the history of the Association.

Meanwhile Halford Mackinder, who had become an MP in 1910, retained a Readership in Geography at the London School of Economics and continued his pioneer work for geography. He was then approaching the height of his remarkable career: Dr Hilda Ormsby paints in an obituary notice a picture of his arrival at the School hot from the House of Commons:

He strode, always a few minutes late, up the Gangway of the Great Hall packed with eager and impatient students, to the rostrum, he paused a moment or two to take in the array of maps that hung there, and then turning to his audience, delivered in his sonorous voice, without ever a note, a perfectly argued and presented synthesis.

New branches continued to emerge as the number of geography specialists increased with the spread of county secondary schools: Sheffield is recorded as starting on 26th February 1907, Bristol on 22nd March 1907, Huddersfield 1907, North London 1907, Manchester 15th October 1909, Chester 1911 and Leeds 1911. Concurrently with these developments the Association was working to increase the number of university departments of geography in Great Britain along with the recognition of geography in examinations. The Class 1 Clerkships examination in the Civil Service was expanded to include geography as an option, the University of London made geography a compulsory subject for the Intermediate examination in Economics, and an optional subject for the Intermediate in Arts and Final BA pass degree. Candidates for the BSc Economics were required to take special instruction in geography. These were small but useful steps on the road to the creation of Honours Schools of Geography in the universities—as yet geography specialists could only aspire to Diplomas.

The retirement of Douglas Freshfield from the Presidency in 1911 marks another milestone in the history of the Association. Thereafter the Presidency became an annual appointment and efforts were made to bring in distinguished 'outsiders' with geographic interests to strengthen the standing of Association. Names such as Hilaire Belloc, Sir Thomas Holdich, Sir William Ramsay, Sir Charles Lucas, Lord Robert Cecil and Sir John Russell appear in the presidential list along with eminent academics such as Dr. G. R. Parkin, Professor E. J. Garwood, Dr. J. Scott Keltie, Halford J. Mackinder and Professor Gilbert Murray (see Appendix C for a list of Presidents). The year 1912 is also marked by the beginning of the Gilchrist Scholarship in Geography (for £100) which was awarded for advanced work in the subject. The first award went to W. E. Whitehouse, a geography master from Welshpool, who later became a Lecturer at Aberystwyth. The number of members in the Association continued to grow steadily and in 1912 reached 1000.

At the Annual Meeting at University College London in 1913 a further important step was taken. The Association had until then flourished without a constitution but after 20 years existence it was beginning to acquire financial assets which needed Trustees, whilst the continued growth required formalisation of its government. A constitution was therefore drawn up and formally adopted at the Annual General Meeting on Thursday 9th January 1913: it is a fairly simple document more or less describing the organisation of the Geographical Association as it had evolved up to 1913—the branches were included in the constitution and Life membership (for £3.50!) was introduced. Full details will be found on pages 47–49 of *The Geographical Teacher* No 35 Vol. VII Pt 1 for 1913.

*The First World War*
The Presidency for 1914 went to Dr. J. Scott Keltie (then Secretary of the Royal Geographical Society); this was a very appropriate appointment as it marked the 30th anniversary of his commission to produce the famous Scott Keltie report. Membership grew to 1,144, new branches appeared at Southampton and Birmingham and the Gilchrist Geography Scholarship was awarded to G. E. Joyce, an assistant master in Barnsley. The impact of World War 1 was not felt for some time and Branch reports continue to reveal a high level of activity, but the problems began to show in 1915 when the membership dropped for the first time to 1,107. Disaster eventually struck however with the unexpected death of Professor A. J. Herbertson on 31st July 1915. The sudden removal of the Honorary Secretary and Honorary Editor compounded with the war problems produced a crisis situation. By the beginning of 1916 many branches were curtailing their activities or closing down 'for the duration': significantly the Gilchrist Geography Scholarships were awarded to Miss Dorothy Adams (1915), Miss Catherine E. Clegg (1916) and Miss M. J.

Photo 11. Professor Percy Maude Roxby, 1880–1947. Honorary Associate Editor 1916–1932; President 1933.
Photo: Geographical Association Archives.

Connaughten (1917). *The Geographical Teacher* continued to appear on time however as H. O. Beckit (Oxford) and P. M. Roxby (Liverpool) were persuaded to take over the editorship pending the appointment of a new Honorary Secretary.

The middle years of Herbertson's editorship of *The Geographical Teacher* witnessed a broadening of the topics dealt with in the Journal: we begin to find geographical surveys of sample regions and overseas areas as well as the expected continuing discussion on the teaching of the subject. Cornwall, Cumberland, Dorset, Essex, Lancashire, the Fenland and the London area all figure prominently although perhaps one of the most important contributions in determining the pattern of future accounts is Percy M. Roxby's survey of East Anglia. Overseas, Nigeria, South Africa, Norway, Egypt and New Zealand are considered. The importance of physical geography is stressed with notable contributions from eminent authorities such as Professor Penck, Professor Garwood, and Professor A. P. Brigham. Map projections, map work, the regional approach, the geography laboratory, were other topics given attention.

The broadened approach continues throughout the later years of Herbertson's editorship with new problems such as the teaching of geography and history as a combined subject (H. J. Mackinder) and investigations into the human geography of Britain (H. J. Fleure) appearing. Regional accounts of Western England, the Port of London, Yorkshire, the Weald and Staffordshire were, with the previous accounts, to lay the foundations for subsequent systematic surveys of the regions of Great Britain in the 1920s and 1930s. By the time of Herbertson's death *The Geographical Teacher* was well established and clearly reflected the breadth of the subject, whilst many articles had become prescribed reading in the training of geography teachers. This was to lead in the next editorship to a new venture—the sale of reprints of the articles most in demand.

The death of Professor Herbertson in July 1915 (and that of Mrs. Herbertson a few weeks later) could not have occurred at a worse time for the Geographical Association. Already disorganised to some extent by World War 1 the Association faced further problems with the resignation of Dr. Unstead from the post of Honorary Corresponding Secretary: and all taking place within the Presidency of Hilaire Belloc who was not entirely *au fait* with the geographic academic world. The Council, mainly as a result of James Fairgrieve's action, wisely decided to postpone the election of a new Honorary Secretary but as noted above it did persuade H. O. Beckit (Oxford) and P. M. Roxby (Liverpool) to take on the Editorship of *The Geographical Teacher* for the time being. Miss E. J. Rickard undertook the duties of the Honorary Correspondence Secretary and the Association struggled on to 1916 when H. J. Mackinder was elected President and new life

was breathed into the organisation.

Despite the war geographical progress was being made in a quiet way: an endowment was given to the University of Liverpool in 1916 which enabled it to create a Chair in Geography for Percy M. Roxby, and within a short space of time another endowment made to the University College of Wales at Aberystwyth permitted the establishment in 1917 of a Chair of Geography and Anthropology for H. J. Fleure, who had been lecturing in Aberystwyth on zoology, geology, botany and geography combined with a research interest in anthropology.

As a result of a recommendation from James Fairgrieve (Fig. 3), Halford Mackinder, with the Council's backing, approached Professor Fleure to fill the post of Honorary Secretary and so initiated in 1917 the fourth phase of the Association's history: this correlates with the rise in eminence of Aberystwyth geography and the relative decline of Oxford geography— Professor Herbertson's chair was personal and H. O. Beckit succeeded as a Reader; it would be another 17 years before an established chair in geography appeared in Oxford.

Before passing to the fourth phase mention should be made of a number of other marginal activities of the Association promoted largely as a result of A. J. Herbertson's influence. Summer schools at Oxford and Cambridge were a regular feature of the geographical calendar from the beginning of the century and later appeared at Aberystwyth, Sheffield and Edinburgh. The Summer School at Oxford under Herbertson was much sought after and is regularly reported upon in *The Geographical Teacher*. The Geographical Association also supported the movement for Regional Surveys initiated by the British Association for the Advancement of Science, and worked with Patrick Geddes during World War 1 mounting a series of conferences on this theme in London, Ludlow and Newbury (Berks). One can detect in these activities the influence of the French geographer Vidal de la Blache.

Another activity for the teacher's benefit was early co-operation with the Ordnance Survey— the Association devising schemes for supplying quantities of Ordnance Survey maps at educational rates: an arrangement which was also of great potential benefit to the Survey as the teachers were in the process of creating the Survey's future market. This co-operation goes back to 1903 and was one of the earliest formal actions of the Association during Herbertson's Secretaryship.

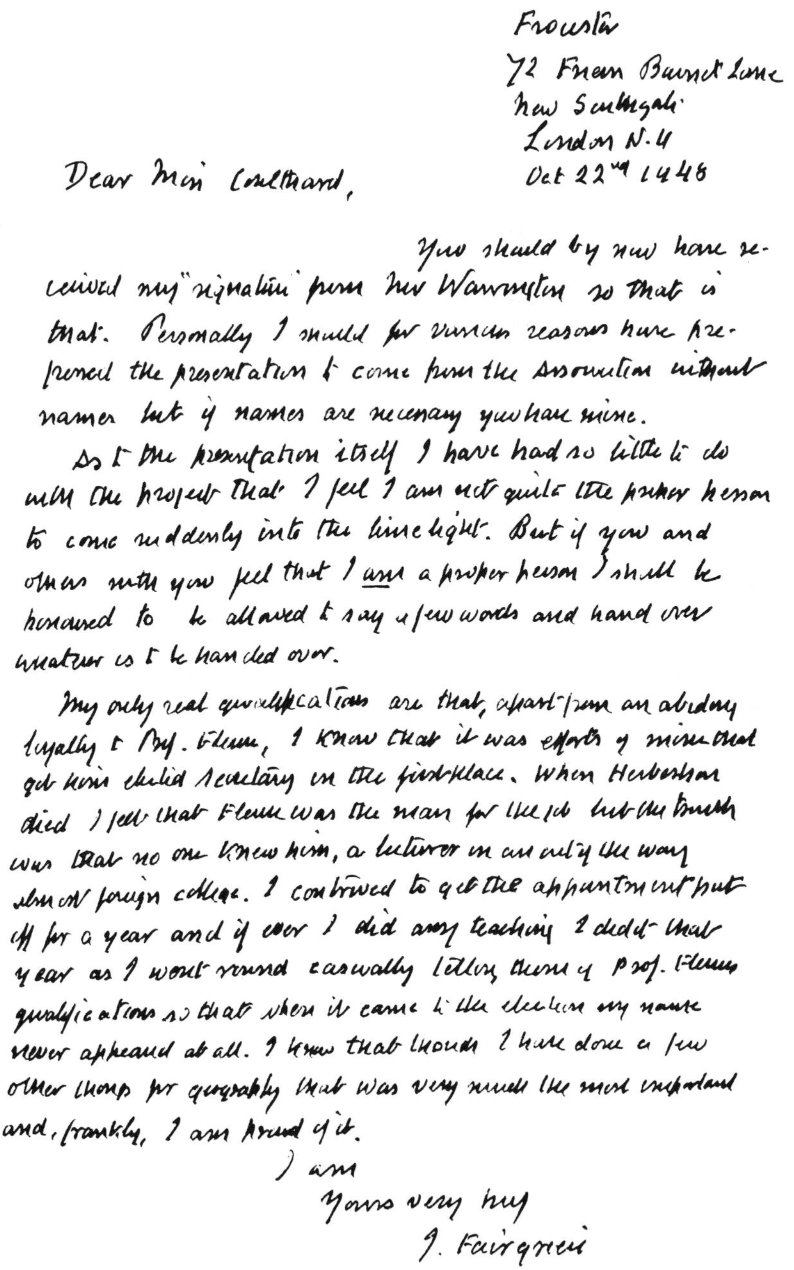
Frouster
72 Friern Barnet Lane
New Southgate
London N.11
Oct 22nd 1948

Dear Miss Coulthard,

You should by now have received my "signature" from Mr Warrington so that is that. Personally I should for various reasons have preferred the presentation to come from the Association without names but if names are necessary you have mine.

As to the presentation itself I have had so little to do with the project that I feel I am not quite the proper person to come suddenly into the limelight. But if you and others with you feel that I am a proper person I shall be honoured to be allowed to say a few words and hand over whatever is to be handed over.

My only real qualifications are that, apart from an abiding loyalty to Prof. Fleure, I know that it was efforts of mine that got him elected Secretary in the first place. When Herbertson died I felt that Fleure was the man for the job but the truth was that no one knew him, a lecturer in an out of the way almost foreign college. I contrived to get the appointment put off for a year and if ever I did any teaching I did it that year as I went round casually letting them know of Prof. Fleure's qualifications so that when it came to the election my name never appeared at all. I know that though I have done a few other things for geography that was very much the most important and, frankly, I am proud of it.

I am
Yours very truly
J. Fairgrieve

Figure 3. Facsimile of the letter from James Fairgrieve to Miss Coulthard which explains how Professor Fleure became Honorary Secretary and Editor of the Association. Geographical Association archives.

# ORDNANCE SURVEY MAPS FOR SCHOOLS.

THE Committee of the Geographical Association recently sent a memorial to the Board of Agriculture asking the Board if they would supply sheets of the one-inch Ordnance Survey maps at a reduced rate when orders for considerable numbers were given for teaching purposes. Mr. Mackinder, representing a Committee of the Royal Geographical Society, and Mr. Graham Wallas, representing the London School Board, also brought the matter before the Board. The Board have received the memorial in the most cordial way, as the following official reply (received as we were going to press) shows :—

SIR,—I am directed by the Board of Agriculture to acquaint you that they have had under their consideration the suggestions made in the memorial of the Committee of the Geographical Association, which accompanied your letter of the 12th ult. The Board are very much in sympathy with the aims the Association has in view, and they would be prepared to authorize the Ordnance Survey Department to produce and supply to Educational Authorities a special edition of any sheet cr sheets of the outline one-inch map, printed on cheap, but reasonably strong, paper at the following prices :—200 copies, £1. 5s. ; 500 copies, £2 ; 1,000 copies, £3 ; 5,000 copies, £12. For larger numbers the estimated price would be £2 per 1,000 copies. These prices will be found to work out to something less than 1d. a sheet in the case of an edition of 500 copies, and in the case of larger editions to much less. They are based on the actual cost of labour and material, the cost of running machinery, and other indirect charges, with a small margin for contingencies. If two or more sheets were required to be joined, in order to give a map round a convenient centre, the price would be slightly increased.

*The Board would, however, be bound to stipulate that any maps thus supplied should not be sold, and a heading would be printed on the maps to this effect.—* I am, Sir, your obedient servant, A. W. ANSTRUTHER, *Assistant Secretary.*
Hon. Secretary, The Geographical Association.

The thanks of all teachers of geography are due to the Board of Agriculture for so promptly and liberally meeting their wishes and giving them the means of basing the teaching of geography on that of the school district in a way hitherto impossible, owing to the cost of the Ordnance sheets. Every pupil in a school should hereafter possess, and be taught how to use, a copy of the local sheet.

Figure 4. Facsimile extract from *The Geographical teacher*, June 1903, page 83. The first official negotiation between the Association and the Ordnance Survey. The Ordnance Survey was then under the wing of the Board of Agriculture. It should also be remembered that in 1903 there were 240 pennies to the pound.

# The Fourth Phase

## H. J. Fleure as Honorary Secretary
## 1917–1946

### A. Aberystwyth as Headquarters 1917–1930

In 1917, prior to his appointment to the Chair of Geography and Anthropology at Aberystwyth, Herbert John Fleure, on the recommendation of James Fairgrieve, was invited by Halford Mackinder as Chairman of Council to succeed Herbertson as Honorary Secretary. Characteristically Fleure saw this offer as an opportunity 'to promote international understanding' as well as promoting geography. He accepted and the following year also took on the task of Honorary Editor of the Geographical Association's journal and other publications: for the next 30 years he directed the affairs of the Association almost single-handed and with the minimum of administrative help. His name, his work, and his department at Aberystwyth, and subsequently at Manchester, would become known to all geographers world-wide during a vital period in the development of the subject in Great Britain. His work for the Association was immense, establishing and running a central office, building the financial resources from virtual penury to modest security, administering a valuable loan collection of texts, and at the same time advocating at the highest national levels the proper recognition of geography. The Geographical Association initially had no resources for real secretarial help and the Honorary Secretary had to record the minutes of all meetings, discussions, reports, etc., by hand. A comment in an Editorial by Fleure in the summer edition of *The Geographical Teacher* for 1920 is revealing although it has a familiar ring. 'Donations are specially solicited to meet the heavy costs of this work for the promotion of better mutual knowledge among teachers of many lands. The price of printing has risen seriously, paper is almost fabulously dear, postage is increasing seriously, and the slender resources of the Association are being taxed to the uttermost'.

The one piece of assistance which was available in the early days was with the journal. Professor Roxby was persuaded to continue as an Associate Editor and served as such from 1916 to 1932. The slide collection was also handed over to the Diagram Company which concentrated on selling sets to teachers rather than arranging for loans.

Photo 12. Professor Herbert John Fleure FRS, 1877–1969. Honorary Secretary 1917–1946; Chairman of Council 1948–1969; President 1948.
© Hull Daily Mail.

19

The World War 1 years were undoubtedly formative years for geography in Great Britain. There was an increased public interest in the subject as a result of world-wide military and political operations. The importance of geographical knowledge for politicians, statesmen and the press was only too obvious and the effect of actual geographical conditions on military strategy and tactics was clearly seen. But when it came to educational matters geography was still relatively poorly represented. The Two Cultures, Arts/Science, tradition has always bedevilled the position of geography in the educational world and impeded its progress as it tends to be seen 'in the other camp' by those secure in their own two educational enclosures. We find this situation present in World War 1 when the inevitable committees to consider postwar developments began to emerge.

The Board of Scientific Studies and the Council of Humanistic Studies were set up by the Government and geography had to argue for a place on both in order to promote its bridge character and integrating philosophy. The archives of the Association and the Royal Geographical Society contain substantial files on these committees indicating all the work that was going on 'behind the scenes' to establish geography in the schools and universities. One key element in the movement was the Board of Education which controlled much of the then examination structure. Although at an elementary level geography had achieved examination status in some structures and was beginning to obtain a foothold in the universities with various diplomas, it did not appear in any of the school Higher School Certificate groups of subjects. When the omission was pointed out to the Board of Education the reply received was that geography was not included as no students offered it for examination! However the Board later agreed to consider geographical syllabuses put forward by individual schools and this eventually led to the appearance of Higher School Certificate examinations in geography in the early 1920s.

These problems span the later months of Herbertson's secretaryship, the interregnum, and the early years of Fleure's appointment. Other significant events also took place with the change of the Honorary Secretary. We see the creation of Standing Committees for the first time, although these bear no resemblance to the present committee structure of the Association. At the Annual General Meeting in January 1918 four standing committees were created to deal with Exhibitions (Chairman, Leonard Brooks), Regional Surveys (Chairman, H. E. J. Peake) Syllabuses and Examinations (Chairman, initially J. F. Unstead but subsequently J. Fairgrieve) and Books, Maps and Atlases (Chairman, initially J. Fairgrieve but subsequently Miss L. M. Hardy). At the same time the Constitution was revised to take account of the changes in the allocation of duties between the Honorary Officers.

Although World War 1 was still in progress 1918 saw a dramatic upturn in the fortunes of the Association, much no doubt related to the improved public perception of geography combined with the promotional work of the Association. By the end of the year the membership had leapt back to 1458 (from 996 in 1917), nearly all branches were operating, and new ones had appeared in Central Lancashire (Preston and Blackpool), North Lancashire, Plymouth, Malvern, Wigan, Cardiff and North Wales (Bangor) whilst the Irish Geographical Association became affiliated to the Association. The country was gradually being covered with a network of branches enabling geography teachers, even in relatively remote areas, to meet colleagues from time to time. Despite the continuing war there was an air of optimism with all this progress, which even expressed itself in the appearance of preliminary plans for study tours and field work in France as soon as hostilities ceased. The Board of Education even ran a Summer School in Aberystwyth in August 1918, possibly through Fleure's influence, to discuss problems of geographical teaching. The first Herbertson Memorial lecture was given in Oxford and repeated in Liverpool in November 1918 by Professor Franz Schrader of Paris on 'The Foundations of Geography in the Twentieth Century'.

## The Inter-War Years

The ending of World War 1 in November 1918 released a burst of educational activity which, combined with the missionary work of the previous years, had a profound effect upon the Geographical Association and the recognition being given to geography nationally. The membership of the Association leapt year by year from 1458 in 1918 to 2379 in 1919, and from 3965 in 1920 to 4159 in 1921. During 1920, under the Presidency of Sir Charles Lucas the Geographical Association 'went international' and invited membership from, in particular, overseas territories of the British Empire, with the result that members were recruited from Accra, Lagos, Nairobi, South Africa, Egypt, Madras, Burma, Ceylon, Straits Settlements, Hong Kong, Canada, Bahamas, Trinidad, Leeward Islands, Australia, Tasmania and New Zealand, as well as China, Japan, USA, Chile and Argentina. One of the objectives of this overseas recruitment was not only membership as such but also the hope of securing authoritative first hand geographical contributions for the Journal.

At home new branches were formed at Bishop Auckland, Crewe, East Suffolk, Essex, Exeter, Hereford, Lincoln, Leicester, Nottingham, Swansea, Thanet, Tottenham, York, and revived at Bristol and Plymouth. The records further indicate that much good work in the cause of geography was undertaken with the local authorities. The surge of activity continued through 1921 with several new branches appearing in London as well as the Isle of Man, Canada, West Africa and Ceylon, although this is said to have been a difficult year because of a national coal strike.

The increasing demand for qualified geography teachers also led in the post-war period to the establishment of Honours degrees in Cambridge, London, Liverpool, Leeds and Wales, together with the organisation of numerous summer schools—at least ten were held in university centres in Great Britain in the summer of 1920 and a further six are reported for 1921. The Government Board of Education itself organised summer schools for geography at Durham and Southampton in 1920 and began giving grants to teachers who opted to take a year off to study for one of the university Diplomas in Geography.

Meanwhile the campaign to get geography fully recognised by the Board of Education, at the Higher Certificate of Education level, continued. A number of eminent individuals had been active in this cause, notably Sir Charles P. Lucas, Professor J. N. L. Myres and Professor H. J. Fleure: this campaign eventually proved successful in the autumn of 1922 when the regulations for the Higher Certificate of Education courses included in paragraph 48E the addition 'Geography combined with two other subjects approved by the Board of which one must be History or a Science'. This catered for both the Arts and Science based students and effectively recognised the bridge nature of geography in relation to the 'Two Cultures'.

Other innovations during this post-war period included the appearance of a 'Spring Meeting' in Southampton in April 1921 by invitation of the Portsmouth, Southampton and Bournemouth branches: this grew out of a Spring meeting of Council. Its success as a result of top level support by, and contributions from, Sir Halford Mackinder and Colonel Sir Charles Close (then Director General of the Ordnance Survey) eventually led to the organisation of an annual spring conference migrating around the regions.

Another interesting development in 1921 was the initiation of "conducted motor tours" organised by Mr. E. E. Lupton of Bradford and led by Mr. C. B. Fawcett of the University of Leeds. These proved so successful that we find a 'Touring Branch' of the Geographical Association appearing in 1922 with a programme of five home tours in Great Britain and two continental tours in France and Switzerland. All were rapidly fully booked: the Association can therefore claim to some extent to be a precursor of the package holiday. The cost of the tours however was somewhat different—the inclusive charge for transport, first class hotel accommodation, all meals and gratuities for 10 days in Switzerland was £15!

A further development in 1921 was the promotion of monograph publication: this began with the appearance of *The Wealden Iron Industry* by Miss M. C. Delany, MA, but the financial risk was undertaken by the publisher, Ernest Benn Ltd., and not by the Geographical Association. Further monographs appeared but the initiative seems to have come more from the publisher than the Association which played the role of an outlet at reduced prices for its members. Branch finances were similarly kept strictly separate from those of the Association, the annual income/expenditure level of which had now reached around £1,250.

The widened interests of the Geographical Association during the initial years of the Fleure era are reflected in the contents of *The Geographical Teacher*. Whilst contributions on the teaching of geography, syllabus content, regional approaches, field work, school journeys, laboratory work and practical activity continue to appear there is a marked increase in articles on overseas locations: all continents and most countries are represented and there is a noticeable emphasis on human aspects. Great Britain is not neglected however as we find interspersed with the overseas articles a continuation of the earlier regional accounts. One important and popular service provided by the journal was the Review Section. An increasing number of geographical books, pamphlets and teaching aids were appearing, and guidance as to their value became critical for the growing band of partly trained geography teachers.

Reading through the Geographical Association's journal at this period one senses that a world consciousness was emerging in higher education, based upon an appreciation of world geography, doubtless helped by the new speed of communication brought about by wireless telegraphy, mass newspapers, the internal combustion engine, improved rail links and the beginning of aeronautical connections. Furthermore geography played a major role in the peace treaty discussions and in the delineation of boundaries after the war: this period was also the heyday of Wilsonism and the League of Nations, and in educational circles the main thought was in the direction of teaching for international understanding and goodwill. Geography had an obvious and significant part to play in this movement and the sudden public realisation of its importance is easy to understand.

The Geographical Association's policy of inviting as President eminent authorities whose activities touched upon geography continued throughout the 1920s and in the first half Sir Charles Lucas, Professor Gilbert Murray, Lord Robert Cecil, Sir John Russell, Professor Sir R. A. Gregory and Professor Sir John L. Myres occupied this position. Of these Sir John Russell and Sir John Myres were to prove long term friends of the Association, exercising considerable influence on the development of the subject. (See Appendix K.)

The number of members continued to rise steadily from 4159 in 1921 to 4462 in 1922 and 4510 in 1923, reaching a peak of 4610 in 1924. The number of branches had reached 67 by 1922 and then tended to remain steady at around this number: inevitably some declined as enthusiastic founder secretaries reached retirement or died, but new branches arose in other places and the map of Great Britain continued to be well covered with the branch network. With each branch aiming at monthly lecture meetings throughout the winter session at least 400 meetings promoting geography were being held nationwide each year and this provided a valuable base for the campaign to gain recognition for the subject. The first stage of this effort had been achieved within the university structure by 1923 at which time all universities had at least one lecturer in geography: the next step, in some ways more difficult, was to promote the creation of Professorships and Departments.

Branches were of course encouraged to develop their own activities and local and regional surveys based on field work were prominent. This activity found a worthy champion in Sir John Russell who took over, after his Presidency in 1923, the chairmanship of the Association's Regional Surveys Committee. This work seems to have inspired the production of the classic book *Great Britain: Essays in Regional Geography* prepared for the twelfth international Geographical Union meeting in Cambridge in 1928. Sir John Russell contributed a lengthy introduction to this notable work of 26 authors edited by A. G. Ogilvie of Edinburgh. This holistic integrated approach to the geography of Great Britain very much reflects the philosophy of the then small band of university geographers.

As soon as the Geographical Association had settled down in Aberystwyth efforts were made by Professor Fleure to provide an improved and augmented postal book borrowing service. Initially the limited collection of books had remained in Oxford in the care of H. O. Beckit but in 1923 the collection was moved to Headquarters in Aberystwyth and the book borrowers were asked to subscribe for the privilege—the additional subscription of five shillings per annum being ploughed back into the acquisition of further books. At first only 500 books were available for loan and as 360 members joined the scheme the shelves were more often empty than full. By 1923 the numbers had risen to 644 books and 500 readers, and in 1924 750 volumes and 533 readers. In 1925 however the Carnegie Trust for the United Kingdom gave the Association a grant of £1,000 and a collection of books for the library. As a result the number of books leapt to over 2,000 and the library subscribers rose to 616. At the same time efforts were made to reinstate the loan collection of lantern slides—the original scheme having lapsed during World War 1. A new modest collection of slides became available from January 1924 derived in the main from travelling geographers who were prepared to donate, with notes, representative sets of slides to illustrate physical and human conditions in overseas territories. By 1926 over 2,000 slides had been assembled: these were in considerable demand at a time when visual aids were very limited. For home areas the Association also suddenly discovered a further source of slides in the railway companies: these maintained for publicity purposes useful collections which could be borrowed. (One sometimes wonders what our early pioneers would think if they could see the present day plethora of audio-visual aids, photographs, film strips, films, transparencies, videos, TVs, recorders and computer displays etc., all in colour!)

The Touring Branch continued with its successful programmes through 1923, 1924 and 1925 offering three or four home tours and two or three continental tours each year. Scotland and the Lake District were favoured at home; Switzerland, the Paris Basin and the Rhine Valley figured abroad.

The Geographical Association concentrated its other main activities on the annual Conference which was organised by V. C. Spary at the London School of Economics each January, and the Spring meeting which after the successful Southampton venture in 1921 became the responsibility of the branches in Leeds (1922), Shrewsbury (1923), Exeter (1924), Reading (1925) and Bristol (1926).

At the policy level the 1922 agreement with the Board of Education on the Higher Certificate of Education teaching in schools was backed up with further recognition of Geography in Science in the External degrees of the University of London; recognition was also obtained in Honours degrees in the University of Manchester, and Arts and Science degrees in the University of Bristol. Shortly thereafter geography was further established in Birmingham, Durham, Leicester and Sheffield. In 1924 the Association was also instrumental in persuading the Ordnance Survey to continue with the quarter sheet 6 inch maps which it was proposing to abandon in favour of composite sheets.

The success on the university front was doubtless one of the reasons why the Royal Geographical Society ceased active participation in the campaign. Although the Society had concentrated its financial effort upon Oxford and Cambridge, financial help had also been given to Manchester, Edinburgh and Wales: during 1888 to 1923, over a period of 35 years, the Royal Geographical Society contributed £7,500 to Cambridge, £11,000 to Oxford, and £2,500 to the other universities to promote geography. By 1924 over £20,000 (old currency) of the Society's resources had been invested in educational activities—representing well over £1,000,000 in present day values.

In the mid 1920s the contents of *The Geographical Teacher* are marked by an increase in the number of articles dealing with overseas locations. At least half of the contributions cover aspects of world geography in which all the continents and most countries are represented. These accounts must have been very valuable for teachers with the limited material then available. There is also the sudden appearance of numerous contributions dealing with climate and the climatic factor, accompanied by tables of climatic data: however, few topics in geomorphology receive attention. The discussion on regional geography and regional surveys continued unabated and doubtless reflects the activities of the Standing Committee on Regional Surveys under the Chairmanship of Sir John Russell. Attention was also given to practical geography, map projections, Ordnance Survey maps, the school geography laboratory, place names and population trends. The Review section was enlarged and lists of articles of geographic interest in other journals published. There is little doubt that the Association's journal was by now providing a useful service for all those engaged in teaching geography.

The rapid growth in the number of branches and the increasing diversity of activities by the Geographical Association necessitated a second revision of the Constitution in the 1920s. The process began in 1923 so that members could consider the proposed changes in 1924 with a view to confirmation in 1925. The revised Constitution covered the increased number of branches, the book and slide arrangements, the Touring Branch and also a new Regional Surveys committee: most important however was the introduction of a smaller Executive Committee, the Council having become too large for the day to day running of the Association.

The mid 1920s were difficult years politically and economically for the United Kingdom and to some extent this is reflected in the Geographical Association's history. The dramatic increase in the membership which had been characteristic of the immediate post-war years came to a halt and although the membership was maintained above 4,000 for the next decade the total fluctuated from year to year: losses in overseas membership partly explains the variations. There were also changes in the Association's long standing officers: the Rev. James Gow, who had been a Trustee since 1914, resigned in 1922 owing to illness (he died in 1923) and was replaced by Principal J. H. Davies of Bangor, but he died in 1926 and was replaced in 1927 by Professor Sir John Myres. E. F. Elton who had been Honorary Treasurer since 1908 died in 1925 and was replaced by Col. Sir Henry Lyons FRS in 1926. Another appointment at this time, although outside the Association, but which was ultimately to prove very significant for the Association, was the addition in 1925 of an assistant, Miss Alice Garnett, to Dr. R. N. Rudmose Brown at Sheffield University.

Other signs of the changing times as well as the adoption of the new Constitution in 1925 included a parallel move to change the name of *The Geographical Teacher* to *Geography* and produce a quarterly issue in place of three volumes a year—moves which were achieved with

some resistance in 1927 and 1929 respectively. Perhaps the most dramatic change in the mid 1920s however was the collapse of the Touring Branch and the sudden cessation of this worthwhile activity. The demise was brought about by a variety of factors, largely because of the fear that the Association might lose its charitable status if the Income Tax authorities regarded the Touring Branch as a business activity, and partly because of unfortunate difficulties being experienced at this time with the Royal Geographical Society.

Relations between the two societies inevitably had to be conducted by the two secretaries and Arthur Hinks, Secretary of the Royal Geographical Society, was the antithesis of Professor Fleure. Hinks' background was 'physical'—mathematics, astronomy, survey and cartography—and his dominance of the Royal Geographical Society was legendary: the epitome of a Victorian, he believed in getting things changed diplomatically, which usually meant over a club lunch in Pall Mall. Fleure's background on the other hand was 'human'—anthropology, zoology, botany and biogeographical—and along with the Association he sought direct change by submissions to the appropriate authorities. Hinks as Secretary of the 'Senior' Society clearly resented the success of the upstart 'Junior' Association and doubtless this was another reason for the cessation of the Royal Geographical Society's interest in educational matters after 1923. The real explosion however came as a result of an unfortunate printer's error in a Touring Branch prospectus which announced the summer's programme as being under the auspices of 'The Geographical Society' instead of the 'Geographical Association'. A copy of the misprinted leaflet fell into Hinks' hands and although it had only been seen by a handful of Association members his reaction was both immediate and furious. With his Victorian values Hinks strongly objected to any commercialisation of the Royal Geographical Society and he stirred up a real storm in a teacup over the misprint, eventually forcing the Association to make a public apology in *The Times*. One cannot but wonder what Hinks' comments would be on recent activities of the Society!

Despite the economic problems of the mid 1920s the second half of the decade was marked by consolidation and stability in the Geographical Association's history. The membership fluctuated slightly, ranging from 4351 in 1926, to 4447 in 1927, 4449 in 1928, 4345 in 1929 and 4233 in 1930. The Library continued to grow from around 2000 volumes in 1926 to nearly 4000 in 1930 with the number of borrowers rising to over 1000. The income/expenditure level fluctuated around £1500 per annum. New branches appeared at Oxford, Cambridge, Derby, Oldham, Peterborough and Stafford whilst lapsed branches were revived at Huddersfield, Mid-Glamorgan and Chester. The policy of inviting distinguished individuals whose interests touched upon geography to fill the presidential post continued: the position being occupied by the Rt. Hon. W. A. Ormsby Gore (Lord Harlech) in 1926, Col. Sir Charles Close in 1927, Dr. Vaughan Cornish in 1928 and Col. Sir Henry G. Lyons in 1929. (See Appendix K.) A new pattern however emerges in the 1930s when a tendency to acknowledge the services of the founders can be seen.

The Annual and Spring Conferences had by now become well established. The Annual Conference was held during the Christmas vacation at the London School of Economics, being organised by Mr. and Mrs. V. C. Spary, who achieved renown for the geographical games and entertainments they provided at the annual dinner. The Annual Conference was also marked by a growing exhibition of geographical books and apparatus. The Spring Conference took place in the Easter vacation as result of invitations from branches in Bristol (1926), Liverpool (1927), Oxford (1928), Hull (1929) and Birmingham (1930).

The revised constitution of 1925 gave rise to a new committee structure in 1926 and 1927, providing the basis for what would become the long term structure. In 1926 a Standing Committee under Mr. C. C. Carter of Marlborough College was created for the development of higher geographical teaching and this was followed in 1927 by Standing Committees for Primary Schools under the chairmanship of Mrs. Katz, and Secondary Schools under the chairmanship of Mr. C. B. Thurston. The Standing Committee for Syllabuses and Examinations continued, whilst the Standing Committee for Regional Surveys under the chairmanship of Sir John Russell

became increasingly active when it acquired Dr. L. Dudley Stamp as its Honorary Secretary: a by-product of the interest stimulated in regional surveys was seen in the special publication of *Great Britain: Essays in Regional Geography* prepared for the 1928 Cambridge meeting of the International Geographical Union. Of greater subsequent significance however was the independent appearance in 1929 of a Land Utilisation Survey map of the County of Northampton published by the Ordnance Survey as a result of the initiative of a geography teacher. Mr. E. E. Field, who, with the support of the Northamptonshire Education Committee operating through its Director Mr. J. L. Holland, had organised a field survey based on the six inch maps.

Dr. Stamp was quick to realise the potential value of a national Land Utilisation Survey for Great Britain as a whole and immediately proposed through the Association committee the establishment of a central organisation at the London School of Economics with representatives from the Geographical Association, the Ordnance Survey, the Ministry of Agriculture and the Education Authorities, to undertake the task. The financing of such an operation was however clearly outside the Association's remit and in 1930 the Land Utilisation Survey

Photo 13. Professor Sir L. Dudley Stamp CBE, 1898–1966. Honorary Treasurer and Trustee 1955–1966; President 1950.
Photo: Bassano and Vandyk. Royal Geographical Society.

of Britain hived off from the Association with an organisation of its own under Dr. Dudley Stamp as the Director. Thus emerged one of the most notable achievements of inter-war geography in Great Britain.

Although the Touring Branch had collapsed the Geographical Association continued its interest in the active support of organisations such as the Le Play Society, which was mounting regular visits to the continent. Horizons were widened even further in August 1929 and August 1930 when trans-Atlantic North American tours by Cunard liners were organised privately.

In the universities geography continued to gain strength and recognition. Ll. Rodwell Jones was appointed Professor at the London School of Economics in 1925, O. H. T. Rishbeth achieved a personal chair at Southampton and W. W. Jervis became a Reader at Bristol in 1926. Frank Debenham became a Reader in Cambridge in 1927 and C. B. Fawcett a Professor at University College London in 1928. Another significant breakthrough came in 1928 when Alfred Steers persuaded St. Catharine's College Cambridge to institute a Scholarship in Geography, first awarded in 1929. This move was to have profound effects on the development of the subject in later years (Balchin, 1989).

In 1927 *The Geographical Teacher* had become *Geography* and in 1929 publication increased from three to four times a year. The contents during this period continued the pattern of the early 1920s with the emphasis on overseas locations for many contributions, although we find home interests well represented. Climatic and regional themes remain prominent and urban surveys begin to appear. The journal also carried a useful section on teaching problems as well as an augmented review section. Alongside the journal the Association also tried to promote research publications: two were issued as supplements— W. Fitzgerald's *Historical Geography of Ireland* which appeared in 1926 and Miss E. Simkins' *Agricultural Geography of the Deccan Plateau* issued in 1927, but these sold very slowly and it was to be some time before further experimental publication was attempted.

During this period the Association suffered the loss by death of Sir John Keltie in 1927:

although subsequent to his famous report he spent most of his working life as Editor of the *Statesman's Year Book* he maintained his interest in geography and lived long enough to see most of the recommendations in his report come to fruition.

### B. Manchester as Headquarters 1930–1946

Another watershed in the history of the Geographical Association came in 1930 when Professor Fleure was invited to take up the newly established Chair of Geography in the University of Manchester. Happily Fleure had no wish to relinquish his Honorary Secretaryship of the Association but this then raised the question of the location of the Association's Headquarters, complicated on this occasion by the existence of a library of some 4,000 books and the problem of a very efficient Assistant Secretary, both located in Aberystwyth. Dr. Spurley Hey, Chief Education Officer in Manchester, solved the accommodation problem by offering the Association free office and library space in the Manchester High School of Commerce and this was gratefully accepted. Sadly Miss R. M. Fleming, who had so ably supported Professor Fleure as a paid assistant secretary for many years, preferred to remain in Aberystwyth: her services to the advancement of knowledge and education were recognised in 1935 by the award of an honorary degree of *Magister in Scientia* from the University of Wales, and a place on the Civil Pension list. Her successor in Manchester was Miss M. E. Owen, who had had experience of the Aberystwyth office. Once again the Association was fortunate as Miss Owen proved to have a strong business sense, a zeal for the library, and great skill in proof correction and publishing matters. At the same time the printing of the journal and other publications was transferred from the Montgomeryshire Press to Percy Brothers of Manchester with whom the Association was to develop a long lasting and close relationship.

The decade of the 1930s proved to be a difficult time for a variety of reasons. The national economic recession and the international political atmosphere of the second half of the inter-war period led to a collapse of the optimism of the 1920s, although the adverse effect on the membership numbers was probably more related to a decline in the birthrate after 1921 and also a reorganisation of educational facilities which turned many all-purpose all-age schools into junior schools up to age 11 only. These factors combined to reduce the need for geography specialists. Although membership was held at around the 4,000 mark until 1935, thereafter there was a steady decline to 3646 on the eve of World War 2.

One notable feature of this period was a change of policy in the selection of Presidents for the Geographical Association. Although eminent authorities still figure the Association had by now become so well established it was able to introduce into the Presidential list some of its founder and inner circle members from time to time. B. B. Dickinson was so honoured in 1930, Dr. H. R. Mill in 1932, Professor P. M. Roxby in 1933, James Fairgrieve in 1935 and C. C. Carter in 1939. Other Presidents and good friends of the Association included Sir Leslie Mackenzie (1931), Lord Meston (1934), Sir Josiah (Lord) Stamp (1936), Professor Sir Patrick Abercrombie (1937) and Sir Thomas Holland (1938). (See Appendix K.)

Despite the membership problem the work of the Geographical Association continued in the by now well established format of the Annual Conference, Spring Conference, publication of *Geography*, Branch activities and the work of the Standing Committees. The Annual Conference (see Fig. 5) with its important exhibition of books and apparatus was by now firmly in place at the London School of Economics each Christmas vacation, but changes took place in the Honorary Conference Organiser. After a notable decade under the guidance of V. C. Spary the mantle of office fell in 1933 on Dr. H. J. Wood for two years and then passed to S. H. Beaver, both of the Joint LSE-King's College School of Geography. The migratory Spring Conference with its emphasis on field excursions attracted substantial numbers each year but the need to provide inexpensive residential accommodation restricted locations to university centres with halls of residence. The sequence in the first half of the decade was Birmingham (1930), Manchester (1931), London (1932), Liverpool (1933), Glasgow (1934), Nottingham (1935), and Sheffield

(1936). The apparent anomaly for 1932 is explained by the fact that the 1932 Spring Conference was originally scheduled for Heidelburg but the arrangements had to be abandoned because of currency and political problems in Germany.

Branch locations had by now stabilised out at around the 55 mark, with activities concentrating upon a regular lecture programme in the winter months producing some 350 meetings countrywide, often backed up with field excursions, and being led in this aspect by the Bradford branch which achieved renown by taking 890 pupils to York and Whitby on a day excursion in 1933!

The improved facilities for the Library in Manchester enabled the book collection to become more of a reference library and use by the members increased steadily despite the fluctuating membership. The increased use was in part related to the appointment of T. C.

Photo 14. Thomas Cotterill Warrington 1869–1963. Honorary Librarian and Member of Council 1930–1954; President 1942–1945.

Photo: Geographical Association archives.

# ANNUAL CONFERENCE OF THE GEOGRAPHICAL ASSOCIATION.

---

TO BE HELD IN THE LONDON SCHOOL OF ECONOMICS, HOUGHTON STREET, ALDWYCH, LONDON, W.C. 2, FROM WEDNESDAY, 2ND JANUARY, TO FRIDAY, 4TH JANUARY, 1935, AND EXCURSIONS ON SATURDAY, 5TH JANUARY, 1935.

---

A PUBLISHERS' EXHIBITION of books, maps and appliances for the study of geography will be open on Wednesday, Thursday, and Friday, and members are especially asked to visit it.

---

### TUESDAY, 1st JANUARY, 1935.

11·0  a.m.  Meeting of the Public and Preparatory Schools Group at the Royal Geographical Society's house in Kensington Gore, London, S.W. 7.

### WEDNESDAY, 2ND JANUARY, 1935.

10·0  a.m.  Opening of Publishers' Exhibition and short address by Mr. Ellis W. Heaton.

10·15 a.m.  Council Meeting.

11·45 a.m.  Presidential Address by the Rt. Hon. Baron Meston of Agra and Dunottar on " The Geography of an Indian Village."

2·0  p.m.  Annual Business Meeting.

3·0  p.m.  " Ordnance Survey History."  Lantern lecture by Brigadier H. St. J. L. Winterbotham, C.M.G., D.S.O.

4·0  p.m.  Tea for members, by kind invitation of the London Branches, in the Refectory.  Admission by ticket only, to be obtained before 2 p.m. on Wednesday, 2nd January, at the G.A. stall.

4·45 p.m.  Meetings for Teachers in Primary Schools, in Secondary Schools, and in Training Colleges to receive reports of their Standing Committees.

5·30 p.m.  Symposium on Russia (in conjunction with the Le Play Society). Short papers by Mr. R. A. Pelham, M.A., Dr. A. S. J. Baster, and Mr. L. Brooks, M.A., to be followed by discussion.  Chairman : Sir John Russell, D.Sc., F.R.S.

### THURSDAY, 3RD JANUARY, 1935.

9·30 a.m.  Publishers' Exhibition.

10·0  a.m.  Lantern Lecture by Dr. L. Dudley Stamp on " Planning the Land for the Future : a Comparison of Land Utilisation Studies in the United States and Britain."  Dr. Stamp will also exhibit a series of maps of various types from the United States.

11·30 a.m.  "The Topographical and Toponymical Aspect of Place-names Study." Lecture by Dr. Allen Mawer, Provost of University College, London.

2·30 p.m.  " Geography and International Problems." Lecture by Dr. G. P. Gooch.

4-30 p.m.  MEETING FOR TEACHERS IN SECONDARY SCHOOLS. " The Equipment
of the Geography Room." Paper by Mr. L. B. Cundall, M.Sc.
Discussion to be opened by Miss E. K. Cook, B.A.

MEETING FOR TEACHERS IN PRIMARY SCHOOLS. Discussion, in
conjunction with the Publishers' Exhibition, on " Junior Text-books."
Discussion to be opened by Mr. E. J. Orford.

7-0  p.m.  Assemble for Dinner in the Women's Common Room ; Dinner at
7-30 p.m. in the Refectory.

### FRIDAY, 4TH JANUARY, 1935.

9-30 a.m.  Publishers' Exhibition.

10-0  a.m.  " Water Supply." Lantern Lecture by Dr. Bernard Smith, F.R.S.

11-30 a.m.  " Some New Geographical Films," introduced by Mr. J. Fairgrieve,
and exhibited at the Gaumont-British Theatre, Shaftesbury Avenue,
W. 1.

2-30 p.m.  Joint Meeting with the Le Play Society Student Group. " The
Stubaital (Austrian Tyrol)." Speaker : Mr. A. E. Moodie. The
chair will be taken by Mr. K. C. Edwards, President of the Student
Group.

4-15 p.m.  MEETING FOR TEACHERS IN TRAINING COLLEGES, in conjunction with
the Geography Branch of the Training College Association. " Schemes
of Work in Geography for Training College Students preparing for
a Teacher's Certificate." Speakers : Miss E. H. Carrier, Mr. J. E.
Daniel, Miss O. Garnett, and Miss D. Wilford, followed by discussion.

Public Meeting organised by the Primary Schools Group in conjunction with the School Journeys Association. " The School
Journey from the Geographical Point of View."

### SATURDAY, 5TH JANUARY, 1935.
#### TWO MINIATURE STUDIES IN LONDON.

Each party will be limited to 20 members. Names should be sent to the
Hon. Conference Organiser (Mr. S. H. Beaver, London School of Economics,
Houghton Street, Aldwych, London, W.C. 2.) before January 3rd.

1. THE LOWER VALLEY OF THE HOLBORN, conducted by Mrs. H. Ormsby, D.Sc.
The party will assemble in Room 504 at the London School of Economics,
at 2-15 p.m., when Dr. Ormsby will give a short introductory talk.
The walk will occupy about two hours.

2. BATTERSEA, conducted by Mrs. E. B. Beaver, B.A.
The party will assemble at 2-0 p.m., in the churchyard of St. Mary's
Parish Church, Battersea, on the south bank of the River Thames.
(Bus No. 19 from Southampton Row, or No. 39 from Charing Cross,
to Church Road, on the south side of Battersea Bridge.)

Further details of both these excursions will be posted in the Entrance
Hall of the London School of Economics during the Conference.

———————

The Le Play Society invites members of the Geographical Association to
attend a lecture by Sir E. John Russell, entitled " The Le Play Society's Visit
to Russia," on Wednesday, January 2nd, 1935, at 8-30 p.m., in the theatre of the
London School of Hygiene and Tropical Medicine, Keppel Street, W.C. 1. The
Chairman will be Professor H. J. Fleure, D.Sc.

The Le Play Society's Dinner is also open to members of the Geographical
Association. It is to be held at College Hall, Malet Street, W.C. 1, at 7-15 p.m.,
on Wednesday, January 2nd, 1935—i.e., immediately before the above lecture.
Tickets, price 2/9 each, may be had from the Le Play Society, 58, Gordon Square,
W.C. 1, until Tuesday, January 1st.

Figure 5. A typical Annual Conference programme from the inter-war period.
Geographical Association archives.

Warrington as Honorary Librarian on his retirement from the Headship of Leek Grammar School. The librarian's office had lapsed after the move from Oxford to Aberystwyth as Miss Fleming had been able to cope with the smaller number of books. With the growth in book numbers the need for an Honorary Librarian was again felt. Concurrent with an annual loan approaching 2000 books there was a slide collection scheme which attracted 200 applications each year.

The Standing Committees of the Association continued their critical work in both submitting and publishing a variety of reports. We find Primary Schools issuing a report on syllabuses in 1930 which achieved a large sale, investigating the supply of Ordnance Survey maps in 1931, working on a book of mapwork exercises in 1932 and publishing a pamphlet on suggestions for teaching local geography in 1933. The Secondary Schools Committee prepared a report on geography for the Chambers of Commerce in 1930, worked on a regional account of the North of England in 1932, published a report on the School Certificate Examination in 1933, and discussed in detail the new examination syllabuses in 1934. The latter led to a memorandum on 'The Place and Value of Geography in the Curriculum of Schools other than those administered under the Elementary Schools Code for pupils under the age of 16' which was submitted to the Board of Education in 1934. The Public Schools Committee prepared a report in 1932 on the Status of Geography in the Common Entrance Examination and worked on syllabus clarification. A new Association *Handbook for Geography Teachers* appeared in 1932 under the direction of Miss D. M. Forsaith.

During this period a new classroom aid appeared in the shape of educational films: spearheaded by James Fairgrieve a special group of voluntary workers investigated large numbers of potential films on geographical topics with a view to their classroom use. Another important activity by the Association led to an arrangement with the Ordnance Survey to supply copies of examination maps after the examination results had been published. These specially printed map extracts were to prove extremely helpful for classroom use and being relatively inexpensive could be regarded as consumable items in the geography budget.

At about the same time the BBC began to develop in earnest educational broadcasts and the Geographical Association became involved in the discussions on the presentation of geographical topics. The BBC had been founded in 1924 and geographical type educational texts soon appeared—the first was on May 9th 1924 when Sir Francis Younghusband spoke on 'Climbing Everest'. The early broadcasts tended to concentrate on Travel Talks and the emphasis was on entertainment with limited geographical content, the talks being designed to interest and excite pupils rather than teach geography. This approach led to many formal and informal discussions between the BBC and the Association, the main characters in the two-way discussions being, not surprisingly, Professor H. J. Fleure, Dr. Dudley Stamp, Eva G. R. Taylor and James Fairgrieve.

The early broadcasts were reviewed in the 'Kent Experiment' of 1927–28 which was based on 20 urban and 42 rural participating schools and this revealed the conflict between presentation techniques. The BBC insisted on first hand experience from its contributors, not all of whom were happy with the new medium. As a result of pressure from the Association the Travel Talks were supplemented in the 1930s by a Regional series approach covering the World in four years. By 1934 1,271 schools were listening to the Travel Talks and 663 had registered for Secondary Geography. The records show that one of the liveliest contributors in the 1930s was David L. Linton—only recently graduated from King's College London. The Association assisted in this experiment by organising detailed assessments. The early broadcasts soon revealed the need for supporting maps, diagrams, photographs and were bedevilled by poor reception in the days of 'cat's whiskers wireless'.

The pioneer work of the Geographical Association in Land Use Survey had of course been hived off in October 1930 in the nation-wide Land Utilisation Survey of Great Britain under Dr. L. Dudley Stamp, but a paternal interest was maintained in this project and much help given to it by the Branch membership. The Six Inch survey progressed rapidly in the urban areas and before long One Inch coloured maps were appearing as a result of an ingenious system of county sponsorships and help from the Local Education Authorities.

The Geographical Association's interest in foreign travel which had been set back somewhat as a result of the 'Lupton affair' with the RGS (see page 24) was resuscitated by a close relationship in the 1930s with the Le Play Society which offered annual excursions to the continent. Joint membership of both organisations was a common feature of this period and many of the excursions were led by geographers, such as Arthur Davies to Yugoslavia and K. C. Edwards to the Austrian Tyrol: whilst A. E. Moodie was Secretary and Sir Halford Mackinder succeeded Sir Patrick Geddes as President of the Le Play Society. (see Fig. 5). At home we find the Primary Schools Committee co-operating with the School Journey Association, whilst Branches of the Association were trying out various travel schemes such as the Oxford tour to Gibraltar in 1932 and the KCL/LSE Joint School Fourteen Day tour of the Southern Uplands of Scotland in 1933.

The Association was pleased to see continued improvement in the status of the subject in the universities despite the difficult economic and political problems of the time. Chairs of Geography were established in Cambridge (F. Debenham) in 1931, Oxford (Kenneth Mason) in 1932, Edinburgh (A. G. Ogilvie) and Sheffield (R. N. Rudmose Brown) in 1931, and in Bristol (W. W. Jervis) in 1933. A Readership was established in Leeds (A. V. Williamson) in 1932, K. C. Edwards became a Senior Lecturer in Nottingham in 1931 and a separate Department of Geography was created there in 1933. a Lectureship appeared in Swansea and Honours courses in both Arts and Science started in Belfast in 1931, whilst Jesus College Oxford followed the example of St. Catharine's College Cambridge by offering a Scholarship in Geography in 1933.

The pages of *Geography* in the first half of the 1930s decade continued to carry a balanced contribution of home and overseas regional topics alongside a wide variety of systematic and specialist articles. Overseas parts of the British Empire, and especially India, naturally appear more frequently, whilst at home urban studies begin to be more noticeable although

there is a good mix of physical, economic, demographic, agricultural and industrial contributions. It is in the systematic and specialist articles however where we find a large expansion taking place over a wide ranging field. Articles on geographical films, broadcasting, map work, weather study, etc., jostle with problems of teaching, water supply, exploration, land use survey, careers and population problems, to name but a few of the aspects investigated, all illustrating the diversity of interest in the subject.

The closing years of the 1930s were marked by the fall in membership already noted: the steady average of just over 4000 members was maintained until 1935 but then drops through 3867 in 1936 to 3666 in 1937, 3646 in 1938 and 3384 in 1939. It was thought at the time that this decline was in part due to the foundation of the Institute of British Geographers in 1933 but the effect of the IBG must have been minimal as its members were largely confined to the university geographers and it was to be many years before the total IBG membership exceeded a hundred. It is much more likely that the drop reflects the reorganisation related to the cyclic fall of the birthrate in the 1920s combined with the deteriorating economic and political situation in Europe and to many teachers' straitened financial circumstances.

Despite this apparent setback the Geographical Association's life remained vigorous. a revitalised library under the guidance of T. C. Warrington increased its stock to over 7,300 volumes and the loans increased to over 1,700 per annum. The slide collection was also widely used. Spring conferences were held in Nottingham (1935), Sheffield (1936), Swansea (1937), Durham (1938) and Leicester (1939): the Annual Conference continued in London under the direction of S. H. Beaver from 1936 onwards and a Summer School was held in Aberdeen in 1937. The Presidents over this period were James Fairgrieve (1935), Sir Josiah Stamp (1936), Professor Patrick Abercrombie (1937), Sir Thomas Holland (1938) and Mr. C. C. Carter (1939). (See Appendix K.)

The Geographical Association's Standing Committees continued with a variety of projects: Primary Schools were concerned with school journeys, the impact of 'the wireless' and aspects of teaching; Secondary Schools considered syllabuses of the Higher Schools Certificate, submitted evidence to the Consultative Committee of the Board of Education, encouraged work on the teaching of local geography and prepared papers for the International Geographical Congress in Amsterdam 1938; Public and Preparatory Schools dealt with problems of the Entrance Scholarship Examination, published a News sheet and considered the School Certificate Examinations; Training Colleges undertook a survey of syllabuses and an investigation into the understanding of geography by pupils of various ages. Other committees dealt with Films, Epidiascopes, and Traditional Housing materials.

Branch programmes were also maintained with lectures and excursions combined with additional local activities such as the preparation of population maps, the viewing of films, and contributions to the land use survey then in full swing—although towards the end of the decade some branches report problems as a result of the 'spread of the wireless'. Extensive co-operation is also reported with local Literary and Philosophical Societies, and Section E of the British Association for the Advancement of Science. Also at this time, as a result of the initiative and enthusiasm of Mr. L. S. Suggate, a scheme to link schools with British cargo vessels was developed and from this emerged the British Ship Adoption Society.

The Association was especially gratified to record in 1936 the election of its Honorary Secretary, Professor H. J. Fleure, to Fellowship of the Royal Society. Although the election was mainly as a result of Professor Fleure's anthropological work it did provide another valuable geographical voice in the Royal Society—ironically Arthur Hinks was the only other FRS in contact with geography.

The Geographical Association also submitted evidence to two important Government committees during this period. In 1935 a Departmental Committee was set up under the chairmanship of Sir J. C. C. Davidson to consider how the effectiveness of the Ordnance Survey could be restored and improved. A small committee consisting of J. Fairgrieve, C. B. Fawcett and T. C. Warrington collected evidence and submitted a report as from the

Geographical Association. Concurrently a selection of six Ordnance Survey map extracts were selected for use in schools and a first edition of 5000 was put on sale through the Association. This rapidly sold out and repeat orders followed in 1937.

Later in 1938 the Association prepared maps and submitted evidence to the Barlow Royal Commission on the 'Geographical Distribution of the Industrial Population'. The maps and evidence subsequently went on display at a public exhibition in the Central Library in Manchester.

*Geography* over this period carried regular reviews of the new Ordnance survey map publications by T. C. Warrington: the content of the journal also begins to show a subtle change of emphasis: whilst there is a continued fairly even balance between home and overseas articles the home based contributions are weighted towards historical aspects, practical aspects disappear almost completely and there is a big expansion in the number of articles dealing with specialist topics such as planning, population, industrial location, visual aids, broadcasting, films, place names, micro-climatology and methodology. One important contribution which was to influence teaching for some time to come was a review of the chief systems of classification of regions of the world current amongst geographers: this appeared in *Geography* in 1937 as a result of the labours of Professors J. L. Myres, P. M. Roxby, J. F. Unstead and Dr. Dudley Stamp.

The finances of the Association hardly change with annual income/expenditure figures of £1,500 and assets of about £4,000, but in 1936 with an eye on the approaching 50th Anniversary of its foundation and bearing in mind the then seven year tax covenant scheme the Association launched a 'Jubilee Fund'. In the event it was to be another 17 rather than 7 years before the appropriate celebrations could be organised.

*B.1. The Second World War*
The Second World War had an immediate and much more serious impact on the life of the Association than the First. The initial threat of air raids in September 1939 resulted in an extensive evacuation of school children from vulnerable urban areas to rural locations: this

migration, which also involved the teachers, decimated the autumn programmes of many Branches of the Association, especially in the south-east. Some Branches managed to struggle on but it was not long before many were posting members that activities would have to be 'suspended for the duration'. The next casualty was the Annual Conference scheduled to take place in London in January 1940. The threat of air raids plus the blackout forced a postponement and eventually a combined Annual-Spring Conference was held in the safer location of Blackpool in March 1940. Sir Cyril Norwood had accepted an invitation to become the 1940 President but with the Association in such difficulty it was decided to postpone taking up the position until 'after the war' and Mr. C. C. Carter was asked to continue in a caretaker status.

The one main activity which did continue as normal was the publication of *Geography* and this became a valuable means of holding the core of the Association membership together as the war proceeded. The Standing Committees also remained active throughout 1939 but thereafter their membership was affected as men and then women were progressively called up for war service: it was to be some time before the work of the Standing Committees could be resumed at full strength. In the interim period the burden of keeping the Association alive fell very much on the elderly and retired members, notably C. C. Carter, T. C. Warrington, J. Fairgrieve, L. Brooks, L. S. Suggate and of course Professor Fleure. Paradoxically the war did provide a stimulus for the completion and publication of the results of the Land Utilisation Survey in which the Association continued to take an interest. This information was found to be of value by the County War Agricultural Executive Committees charged with the task of increasing food production in order to reduce imports as much as possible. Later in 1941 Dr. Dudley Stamp was appointed Vice-Chairman of Lord Justice Scott's Committee on Land Utilisation in Rural Areas.

1940 saw the fall of France, the Dunkirk 'miracle evacuation', the air battle of Britain, and the beginning of the air raids on Britain: with these events came the realisation that the war would be a long drawn out affair with the country effectively under siege conditions. Any hope of a speedy end had to be abandoned and the Association faced the need for a wartime plan. Top priority was given to publishing and distributing *Geography* which maintained an unbroken record although paper rationing caused a progressive reduction in size. Any idea of an Annual Conference in London during January had to be abandoned owing to the continued possibility of air raids combined with the blackout: Blackpool having proved successful in 1940 it was decided to combine the Annual Conference with the Spring Conference and venues in the north and west of the country were sought: hence Edinburgh in April 1941 and Exeter in April 1942. Manchester was the first choice for April 1943 but this fell through as a result of accommodation problems and eventually the Association met in Cambridge in August 1943. Branches were urged to keep going as much as possible and there was some success with this policy as there was a drift back home of many evacuees after the failure of the air raids to bring the country to a halt.

The Geographical Association felt itself very much 'in the front line' with the air raids as in December 1940 the High School of Commerce in Manchester, containing the Association's Library and Offices, was saved from an immense air raid fire only by the courage of the caretaker, Mr. Sim, who brought the flames to a halt by keeping a hose working for several hours. Later Lord Stamp (President 1936 and brother of Dr. L. Dudley Stamp) and his family were killed in an air raid, and the Land Utilisation Survey lost most of its stock and many valuable records in a London air raid. The loss of the Land Use stock was rapidly repaired by reprinting but the dislocation was considerable. Probably the narrowest escape the Association experienced was on the occasion of the combined conference in Exeter in April 1942 which was completed just before the disastrous April 'Baedeker' air raids on the city (see Fig. 6).

The August 1943 Cambridge combined conference was notable in that it marked in a somewhat subdued fashion the 50th anniversary of the founding of the Geographical Association. The celebrations envisaged in 1936 when the Jubilee Fund was initiated

## THE GEOGRAPHICAL ASSOCIATION.

### Programme of the

# CONFERENCE TO BE HELD AT EXETER

from THURSDAY, 9th April, to MONDAY, 13th April, 1942.

Meetings will be held in the Washington Singer Laboratories, Prince of Wales Road, Exeter, unless otherwise stated.

*Hon. Local Secretary :*
Miss B. M. Swainson, M.A., University College, Exeter.

### THURSDAY, 9th APRIL, 1942.

7-0 p.m.    Dinner at Hope Hall.

8-30 p.m.   Council Meeting (at Hope Hall).

### FRIDAY, 10th APRIL, 1942.

9-15 a.m.   General Meeting.  All members invited.

10-15 a.m.  Inaugural Address by Principal John Murray.

11-30 a.m.  Prof. H. J. Fleure : " An Interpretation of France."

2-0 p.m.    Visit to the Cathedral under the guidance of Canon McLaren.

4-45 p.m.   Sir John Cunliffe : " Burma, Past and Present."

8-0 p.m.    Teachers' Group-Meetings, and, if time permits, informal discussion (at Hope Hall).

### SATURDAY, 11th APRIL, 1942.

10-0 a.m.   Prof. W. Stanley Lewis :  " The South West as an Example of Isolation, Stagnation and Rejuvenation."

11-15 a.m.  Mr. Hugh Ruttledge : " The Attack on Everest."

2-30 p.m.   Walk around Exeter under the guidance of Prof. W. Stanley Lewis.

8-0 p.m.    Informal Discussion (at Hope Hall).

### SUNDAY, 12th APRIL, 1942.

It is hoped to arrange an excursion into the country ; details will be available later.

Messrs. A. Wheaton & Co. Ltd. are holding an educational exhibition at their shop, 231-232, High Street, Exeter, during the Conference.

# THE GEOGRAPHICAL ASSOCIATION.

## EXETER MEETING,
### 9th to 13th April, 1942.

Accommodation for a limited number of members is available at Hope Hall, Prince of Wales Road, Exeter, at a charge of £2 10s. inclusive (Thursday to Monday), plus 10 per cent. gratuities.  No ration books or emergency cards will be required.

The following hotels, situated between the stations and the Washington Singer Laboratories, are recommended to members who wish to make their own arrangements for accommodation.  As long a notice as possible should be given to these hotels, as Exeter is a reception area.

The Great Western Hotel : 18/6 per day, full board.  Under normal conditions this hotel will guarantee accommodation.

A very limited amount of accommodation may be available at the following :—

The Imperial Hotel : £1 1s. per day, full board.

The Rougemont Hotel : £1 1s. per day, full board.

Beach Hill Hotel, St. David's Hill : 14/6 per day, full board.

The Osborne Hotel, Queen Street : 14/- per day, full board.

Applications for accommodation at Hope Hall, and for the Sunday excursion, should be sent to Miss B. M. Swainson, University College, Exeter, **by 2nd April.**  Please state whether you are willing to share a room.

Figure 6. The attenuated Annual/Spring Conference programme held at Exeter in April 1942 a few days before the two night Baedeker air raids on the city. The cost of accommodation is now of considerable interest.

could not be held and yet another event had to be postponed until after the cessation of hostilities. Mr. Carter meanwhile had handed over the caretaker presidency to Mr. T. C. Warrington who gave an appropriate historical address: but owing to the paper shortage even this could not be printed in *Geography* until the Diamond Jubilee celebrations in 1953.

Another notable contribution at the Cambridge Conference came from Sir Halford Mackinder still geographically active at the age of 82. Much of what Mackinder wrote in papers in 1937 and 1942 finds an echo in the new National Curriculum for Schools (DES, 1990) but in his Cambridge 1943 contribution (essentially an addendum to a major 1942 address on 'Geography as an Art and a Philosophy' in reply to Principal John Murray who had posed the question 'What is Geography?') we find a thought provoking statement which still needs consideration:

> Has not the time come when we should again base at least the secondary stage of our teaching on a global outlook? Is not the entire planetary surface now for many purposes the only 'natural' region, but with the advantage that, being a closed area, it admits of the accurate comparison of many orders of pattern?
>
> (*Geography*, September 1943)

Although the loss of staff to war service made life problematical for the Geographical Association it was not all gloom and doom on the geographic front. There was a general recognition of the value of geographical training and knowledge in the conduct of the war and some 300 geographers are known to have contributed specialist knowledge in a wide variety of activities ranging from survey and cartography to geomorphology, meteorology, climatology and general intelligence. The Naval Intelligence division 'handbooks' (in effect regional geographies) were almost entirely written by geographers (Balchin, 1987).

Despite the wartime problems there were even advances on the university front: in 1940 the Oxford Colleges announced more provision for geography scholarships at Hertford and Jesus and several other colleges included geography as a subject which could be offered. Chairs were established in 1941 at Reading

(Dr. A. A. Miller) and Newcastle (G. H. J. Daysh) and in 1944 at Leeds (A. V. Williamson), whilst at Oxford E. W. Gilbert became a Reader. Other notable events included Sir Halford J. Mackinder's achievement in 1943 of 50 years connection with the Geographical Association, Sir John Linton Myres' knighthood in 1943, but sadly the death of Colonel Sir Henry E. Lyons in 1944 (see appendix K for details of his remarkable career).

One measure of the reduced activity of the Geographical Association in these difficult years is the income/expenditure account which dropped from around £1,500 per annum pre-war to around £1,000 per annum during the war. At one point a deficit of £50 was incurred—a most unusual state of affairs for the Association. One service in particular was continued—largely due to the devotion of the Honorary Librarian, Mr. T. C. Warrington—this was the postal loan of library books. The number of books and journals in the library continued to grow steadily, approaching 10,000 by the end of the war whilst annual loans were maintained at an average of around 1,500 a year. The loan of the slides however proved more difficult and this service ceased in August 1942. Of less impact for the membership but none the less remarkable for sustainability was the way James Fairgrieve kept the film committee in operation, analysing steadily the educational potentiality of numerous films. Schools broadcasts on the other hand were much affected by war-time restrictions. Supporting pamphlets had to be cut and the need to deny the enemy any useful information meant eliminating all up-to-date accounts. The BBC's policy of first hand experience and 'actuality' also reduced the field of potential contributors. Conversely portable recording apparatus was being much improved as a result of war-time needs and this was to have important post-war repercussions.

The success of the D Day landings and the return of the Allies to the continent in June 1944 led to a wave of optimism in the country that the war would soon be over: the Geographical Association itself was caught up in the prevailing mood and after the quiet success of the combined conference in Cambridge in August 1943 plans were made for a similar event in August 1944. Sadly this proved to be a casualty and had to be postponed owing to the unexpected flying bomb and V rocket attacks on SE England. As a result there was no Association Conference at all in 1944—the dates of the statutory Annual Conference had slid progressively through the year from January to Spring, from Spring to Summer and finally into the following year as custom was re-established with an Annual Conference at the City Literary Institute in London in January 1945. This was repeated in January 1946 when Sir Cyril Norwood finally took up his postponed Presidency.

As already noted the publication of *Geography* continued unbroken throughout the difficult war years but the contents were much reduced owing to paper rationing. All the established features—news, reviews, articles in journals, etc., were maintained in an abbreviated form and an effort was made to provide a balanced collection of home and overseas articles. The overseas contributions continued as 'pure' geography and little attempt was made to introduce political or war orientated themes—with so many academic geographers engaged in the Naval Intelligence Division operation a good deal of discretion had to be exercised as to what was published. The same approach applied to the home articles and we find an emphasis on historical and evolutionary aspects as up-to-date statistics were hard to obtain. Among the specialist articles cartography and climatology figure extensively but there was a complete absence of any geomorphological contributions.

As the war gradually ended the Association faced additional problems. Professor Fleure reached retiring age from Manchester University in September 1944 and he accepted a visiting Professorship at Bowdoin College, Brunswick, Maine, for the 1944–45 session; while Miss M. E. Owen who had been a clerk at Manchester since 1930 (and previous to this had been an assistant to Miss R. M. Fleming in Aberystwyth) also retired in 1944. The immediate problems were solved by re-allocating the Secretary/Editor's duties among the President, Dr. Margaret Davies and Mr. N. V. Scarfe, whilst Miss Owen was replaced by Mrs. Mann.

These emergency arrangements coincided with the gradual return of the country to peacetime conditions: a somewhat long drawn out transition period followed during which personnel were gradually released from the Forces, teaching duties were resumed, and a number of key appointments were made in the universities: Chairs were filled in Liverpool, Birkbeck London, Aberystwyth, and Sheffield.

Partly stimulated by the Education Act of 1944 the Secondary Schools Standing Committee and the Preparatory and Public Schools Committee were re-activated, Branches were restarted and from a wartime low of 2261 in 1941 the membership climbed through 2737 in 1944 to 3191 in 1945 and 4265 in 1946.

*B.2. The Post WW2 Years*
The optimism of 1945 that life would soon be back to 'normal' was soon to evaporate however as the realisation dawned that the country had reached a nadir in its economic life. With its lifeblood exports seriously diminished, many cities and industries seriously damaged, with its Empire moving towards independence and the need to maintain a large European military presence because of the Soviet threat, the return to something approaching normality in Britain was to take ten years at least. Conditions in the immediate post-war Britain were in fact worse than during the war years. Rationing was extended and the country suffered a series of unusual natural disasters, the most notable being the cold winter of 1947, the London smog disaster of 1952 and the East coast floods of 1953. It was against this background that Britain and its constituent Associations struggled to achieve normality.

In the case of the Geographical Association the problems were compounded because of an unusual combination of deaths and retirements among key personnel. Having existed for just over 50 years with few casualties or changes a sudden increase in both was perhaps not surprising: but in quick succession the Association lost in 1947 Professor L. W. Lyde (Trustee 1914–47), Professor P. M. Roxby (President 1933), the Rt. Hon. Sir Halford J. Mackinder (President 1916 and Chairman of Committee and Council 1908–1947), Sir Thomas Holland (President 1938), followed in 1949 by C. C. Carter (President 1939–41) and in 1950 by Dr.

Hugh Robert Mill (President 1932). (See Appendix K.)

Amongst the retirements the most serious loss was of course that of the Honorary Secretary/Editor, Professor H. J. Fleure, who had held both posts with great distinction for 30 years. In fact it was commonly acknowledged in later years that during this period of office the Geographical Association *was* Professor Fleure. With his retirement from the Chair in Manchester and a move to join his family in the south of England however it was obvious that he could no longer effectively run the Association with its headquarters still in Manchester, and reluctantly Fleure resigned from the Honorary Secretary/Editor posts as from the end of 1946. The Association could not afford to move from its free accommodation in Manchester and this factor doubtless affected the choice of the replacement appointments. As from January 1947 Dr. Alice Garnett and Professor David Linton, both of Sheffield University, became the new Honorary Secretary and Honorary Editor respectively. Sheffield was considered to be within easy reach of Manchester. Soon after, another retirement from long active service with the Association took place: James Fairgrieve was forced by ill health to give up but not before passing through a one stick, two stick, wheeled chair, bath-chair and basket-bed sequence of attendance at Executive Committee meetings!

Significant changes were also taking place in other geographical quarters, the most notable being at the Royal Geographical Society where Arthur Hinks died suddenly on April 18th 1945. The new Director/Secretary was Lt. Col. L. P. Kirwan. Aided and abetted by Leonard Brooks, who was an Honorary Secretary of the Society as well as a Trustee of the Association, Kirwan established a new era of cordiality and co-operation between the three geographical societies. In the emergency post-war conditions the House of the Society (which had survived undamaged) was made available both to the Association and the re-established Institute of British Geographers for joint meetings and conferences. For several years after the war the Institute held its annual meetings in London at the Society's House integrating its programme in with the Annual Conference of the Association. It was not until 1949 that the migration of the IBG around the provinces started with a meeting at Oxford. From this time there also dates the tripartite annual lecture and reception which was instituted by Leonard Brooks, and which has continued unbroken to the time of writing.

# The Fifth Phase

## Alice Garnett as Honorary Secretary
## 1947–1967

The new Honorary Secretary, who already had close connections with the Geographical Association, faced an immediate problem as post-war developments in Manchester at the High School of Commerce raised the question of the free accommodation enjoyed by the Association. Sadly the Jubilee Fund had barely reached £1,000 and there was no question of buying suitable premises. Eventually the High School had to be relinquished but Manchester was able to offer alternative temporary space at 103 Princess Street for a limited period. Although theoretically only an hour or so away from Headquarters the location of both the Honorary Secretary and the Honorary Editor in Sheffield proved to be inconvenient and following further pressure from Manchester enquiries were as made as to suitable accommodation in Sheffield. Almost miraculously the Sheffield City Libraries Committee was able to offer, through the good offices of the City Librarian, Mr. P. J. Lamb, free accommodation on an upper floor of the new Park Branch Library in Duke Street, Sheffield 2. The move took place in April 1950—with regret at having to leave so many good friends in Manchester but with a certain amount of excitement as the new accommodation offered the possibility of interesting new developments.

### A. Manchester as Headquarters

The period between the formal end of the war and the move to Sheffield had meanwhile witnessed a number of significant activities as the Geographical Association began to resume its normal rhythm. Sir Cyril Norwood eventually took up his long delayed Presidency in 1946, to be followed by Sir Alexander Carr-Saunders in 1947, Professor H. J. Fleure in 1948, Sir Harry Lindsay in 1949, and Professor L. Dudley Stamp in 1950. Normal Annual Conferences, attracting up to 500 participants,

were resumed, complete with Publishers' Exhibitions, at the London School of Economics in January 1947, with S. H. Beaver as the Honorary Conference Organiser. Spring conferences were held in Sheffield in April 1947, Birmingham in April 1948, Liverpool in April 1949, and as a new venture in Falmouth,

Photo 15. Professor Alice Garnett, 1903–1989. Member of Council 1930–1940; Honorary Secretary 1947–1967; President 1968; Chairman of Council 1970–1973.
Photo: The University of Sheffield.

Photo 16. Professor Sidney William Wooldridge
CBE FRS, 1900–1963. President 1954.
Photo: The Royal Society.

Cornwall, in the Spring of 1950. Between 150 and 200 members attended each of these conferences and they were popular events at a time when travel was difficult owing to petrol rationing. Branches were also gradually reactivated and by 1950 some 58 were back at work with programmes of lectures and field excursions. Although not directly responsible the Association also gave publicity to and encouraged participation in overseas field excursions organised by the Le Play Society (eg. Swedish Lapland, August 1946) and its off-shoot the Geographical Field Group organised by K. C. Edwards in Nottingham (eg. Switzerland August 1947 and Norway, Brittany and Switzerland in the summer of 1948). Summer schools were also organised in Northern Ireland in 1948 and 1949.

Considerable stress was also given during this period to the development of school journeys, field work and field studies, spearheaded by the newly formed Council for the Promotion of Field Studies which was begin-

ning to open a succession of study centres (eg. Flatford Mill, Juniper Hall, Malham Tarn, Dale Fort). Many geographers, notably Professor S. W. Wooldridge, C. C. Fagg and G. E. Hutchings, figured in this new movement.

The Standing Committees began to address themselves to the revision of the Association's Handbooks and catalogues: new editions of the Primary Schools Handbook, the Secondary Schools Handbook, Local Studies, Visual Aids, Library and Film Lists, were all set in train. OS Map Sets were reprinted and several thousand were sold. Attention was also given to the consequences of the revocation of Regulation 18(3) in Circular 140 of the Ministry of Education which theoretically liberated schools for visits to places of interest outside the school boundaries. In actual fact this produced numerous problems of different interpretations, timetable difficulties, and insurance problems. Many geography teachers were glad of the help the Association was able to provide in the interpretation of this relaxation of the regulations.

Membership figures fluctuated somewhat after the post-war recovery increasing from 4265 in 1946 to 5200 in 1947 but dropping back to 4400 in 1948 and 4051 in 1949: the rise and fall relates to the opening and closing of numerous emergency training colleges needed to bring the teaching profession back to strength after wartime losses and accumulated retirements.

Another important post-war task which arose as a result of the changes in the Geographical Association's structure and personnel was a Revision of the Statutes and Standing Orders in which the most significant (and at the time contentious) issue was the raising of the annual subscription to twelve shillings and sixpence. Full details of the revision are published in *Geography* for September 1947. The combination of the variation in membership numbers and the change in the annual subscription had an inevitable effect on the annual revenue account which rose from £1746 in 1946 to £1979 in 1947, £2455 in 1948, and £2423 in 1949.

One topic very much under discussion in the Geographical Association at this time was the problem of Social Studies in schools. This came to the fore as an indirect result of the 1944 Education Act and the introduction of the concept of Secondary Modern schools alongside the traditional Grammar Schools. It was argued that Secondary Modern pupils needed broader based thematic courses, and Social Studies—an amalgam of history, geography, sociology, economics and civics, was much in favour. This produced strong reactions from the traditionalists in all these subjects, not least in geography. An oft quoted and highly significant contribution to the debate came from Professor S. W. Wooldridge in his famous lecture 'On Taking the "Ge" out Geography' given to the Annual Conference in January 1949 and printed in *Geography* for March 1949. The Education Committee of the Royal Geographical Society was also moved to produce a statement, with which the Association concurred, deploring the situation (*Geography*, September 1950, p. 181).

The content of *Geography* under the new Honorary Editor, Professor David Linton, underwent a subtle change during this period. The publication of learned journals was still bedevilled with the continued paper shortage and although the main features of *Geography* were preserved the balance was altered. There was a reduction in the amount of space allocated to overseas articles (hardly any appearing in 1947), but there was an increase in home based contributions with an emphasis upon historical, demographic and settlement aspects.

Significantly, in 1949 geomorphology begins to appear once again. The main change however was a large increase in general articles relating to teaching problems—syllabuses, current affairs, social studies, teaching techniques, field studies, schools broadcasts, etc.—along with a noticeable batch of cartographic contributions—OS maps, map projections, atlases, map reading, air photographs, etc.—along with climatology articles—local climates, daily weather, rainfall dispersion, the cold spell of 1947, etc. Some articles reflect the wartime activities of the contributors (eg. D. L. Linton in the Air Photographic Interpretation Unit, and W. G. V. Balchin and N. Pye in the Hydrographic Department of the Admiralty.)

The major changes which the Association was experiencing were paralleled in the academic world of geography as numerous

Photo 17. Professor David Leslie Linton, 1906–1971. Honorary Editor 1947–1964; President 1964.
Photo: University of Birmingham.

deferred retirements took place, replacements were made, and a number of new chairs were created: altogether seventeen new professors of geography appeared between 1945 and 1950 (see appendix L).

## B. Sheffield as Headquarters

Although in the Geographical Association's history the fifth phase has been dated to the change of Honorary Secretary in 1947, the real change came in 1950 with the move from Manchester to Sheffield. The intervening three years were more of a holding operation for the new Honorary Secretary and Honorary Editor as they gradually adjusted to the changing post-war conditions. The move also resulted in staff changes as Mrs. C. D. Mann, who had given such valuable service in Manchester, had to resign. She was replaced by Miss Marguerita Oughton as Assistant Secretary, supported by two junior assistants. After several months of intense activity moving all the files, records, books and other impedimenta across the Pennines the new Headquarters in Duke Street, Sheffield, were formally opened during the week-end of 13th–14th October 1950. The opening reception was followed by a dinner in the University which was attended by nearly a hundred members and guests: this enabled the Association to give formal thanks to Sheffield Corporation for assistance with accommodation.

With this new look and for the time being adequate staff the Geographical Association began to overcome the post-war problems and prepare plans for the Diamond Jubilee celebration in 1953. The culmination of the celebration programme took place in Sheffield on 26th September 1953 and marks another significant point in the Association's history. The Golden Anniversary of the foundation had fallen in the midst of World War 2 and only a muted acknowledgement was possible in the Cambridge meeting of August 10–14 1943. In 1953 however some 200 members from all over the country were able to assemble at the new Headquarters in Sheffield on 26th September for an initial reception before proceeding to the University for the Jubilee Address by Professor Frank Debenham (a survivor from Scott's last Antarctic expedition of 1910–12 and the immediate past President) which preceded the Diamond Jubilee dinner held in the Stephenson Hall. The dinner once again enabled the thanks of the Association to be conveyed to the civic authorities in Sheffield for their munificent assistance in providing rent free accommodation. The high table at the dinner was marked by a distinguished civic representation supported by a veritable galaxy of senior university geographers. A geographic gathering would not have been complete however without field excursions and these duly followed over the week-end. In all this was a truly memorable occasion which is fully reported in *Geography* for November 1953.

The increasing strength of both the Association and also geography in the schools and universities is reflected at this time in the choice of Presidents. Eminent individuals with marginal interests in the subject no longer figure as the Association now turned exclusively to the rapidly increasing professoriate and its own senior members. Professors L. Dudley Stamp (1950), F. Debenham (1952), S. W. Wooldridge (1954), are interspersed over this

period with Leonard Brooks (1951), Dr. O. J. R. Howarth (1953) and L. S. Suggate (1955).

The pattern of conferences re-established after World War 2 continued. The highlight each year was the Annual Conference with the Publishers' Exhibition held in the London School of Economics over the New Year period and organised from 1951 to 1955 by Dr. W. G. V. Balchin. The Easter vacation each year saw the migratory Spring Conference centred on Falmouth (1950), Hull (1951), Tenby (1952), Lincoln (1953), Exeter (1954) and York (1955) with local branches vieing with the invitations. Summer schools were also revived but these were largely dependent on the enthusiasm of individuals prepared to undertake the administrative and academic chores. Such schools were held at Rhoose, Glamorgan (1952), Pulborough, Sussex (1953), Sistrans, Austria (1954) and Aix-en-Provence, France (1955). Headquarters meanwhile had organised the first international Conference of Geography Teachers in 1951 and after the second international Conference in Holland in 1954 Sheffield agreed to become the Headquarters of the International Union of Associations of Teachers of Geography. With all these activities the Geographical Association was increasingly widening its outlook and from 1950 steadily increasing membership once again after the artificial post-war high/low produced by the expanded teacher training programme. The 1950 total of 2944 reached 3781 by 1955.

Financially however the Association faced difficult problems. By the early 1950s post-war inflation was beginning to upset the finances of all societies dependent on annual subscriptions. In the case of the Geographical Association the situation was further complicated by the gulf between subscriptions and expenditure, only half of the latter being met by members' annual fees. The financial year was changed from January–December to September–August in 1951 but this did little to resolve the problem. There was a great reluctance to increase the subscription, a process which in any case took two years with the constitution as it then stood; but eventually it was agreed that this step would have to be taken and a figure of one guinea was decided upon as from September 1955. The annual income and

Photo 18. An interesting archival photograph taken at the Spring Conference in Hull March 1951. Left to right: H. F. Brown, Secretary Hull Branch; Professor F. Debenham, President 1952; Leonard Brooks, President 1951; Professor H. J. Fleure, President 1948; Herbert King (later Professor) Geography Dept., Hull.

Photo: Geographical Association archives.

expenditure account had risen from £2273 in 1950 to £3491 in 1954 with the assets of the Association then running at around £7000. Bearing in mind that most of the financial transactions at the time were in shillings and pennies rather than in pounds (donations of six old pence for the Association's funds are gratefully recorded!) it will be realised that these figures actually represent a high level of activity.

The Section Committees had by now settled down into a steady work programme. The Secondary Schools published a new Secondary Schools Handbook in 1952 and then turned its attention to the content of school atlases: its Chairman for 25 years, Mr. C. B. Thurston, reluctantly relinquished the office in 1953. The Primary Schools Section revised *Geography in* *the Primary School* in 1953 and then concentrated on the preparation of a publication dealing with Sample Studies. The Public and Preparatory Schools Section opened its membership to women teachers in 1951(!) and organised a sequence of annual meetings in Oxford. The Training College Section was particularly active in the organisation of several one day conferences each year in order to meet the criticism that much of the geography teaching in secondary modern schools was being undertaken by individuals without any geographic training. The Visual Aids Section was responsible for an influential publication in 1954 dealing with 'The Geography Room in the Secondary School'. Data collection and survey work was also being undertaken by the

Urban Spheres Committee and two newly founded committees for Further Education (1953) and Field Studies (1954).

Meanwhile the Executive Committee was much concerned with BBC broadcasts, the publication of wall maps, field studies, and syllabus problems which had arisen with the Associated Examining Board and the Scottish Universities Entrance Examination. There was one notable advance in 1954 when Geography in the Higher Civil Service Examination was brought into line with other subjects and allocated 700 marks.

The pre-war discussions with the BBC were resumed in the post-war period as new problems were emerging. The McNair Report on Teacher Training (1944) had recommended that the use of film and broadcast material should be encouraged in the schools but the increasing spread of the cinema revealed the need for more 'reality' on radio which could only be achieved with sound effects, music and dramatic re-creations. This renewed the debate between teachers, specialists, presenters and broadcasters on the form and function of the broadcast material. The Geographical Association wanted 'real geography' but entertainment dominated the thoughts of the broadcasters. The BBC also maintained that 'scientific geography' was too difficult to handle which meant that emphasis was given to 'human geography' with Travel Talks still dominating. The result was that Secondary Modern Schools used the service quite a lot, although mainly on a switch-on switch-off basis; whilst Secondary Grammar schools were minority users but where the teacher prepared the class for the broadcast and then followed it up with exercise work. Supporting colour pamphlets were reintroduced and these subsequently became quite voluminous.

The Central Council for Educational Broadcasting became the Schools Broadcasting Council and the BBC recruited many more specialists — geographers included. The Geographical Association however was disappointed at the loss of the specialist geography committee and argued for the close involvement of practising geography teachers in programme planning and preparation in an effort to achieve more 'real' geography with less sound effects and background music. One

result was the inclusion of some physical geography in the programmes for 1951–52.

But other problems were appearing: Examination Boards and Syllabuses were proliferating rapidly and before long Comprehensive Schools were being organised: all of which made it difficult to programme for the Secondary level. By 1960 it was clear that a four or six year cycle of world geography was not tenable as changes were taking place too quickly on a multitude of fronts. There was however the realisation that radio could effectively cope with the reporting of rapid change and effectively up-date inaccurate textbooks, whilst 'actuality' could be achieved by on the spot reporting. The 1960 Pilkington Report on Broadcasting therefore supported the educational role of the BBC.

As might be expected *Geography* also experienced changes during this period but improvements could only be introduced slowly owing to the continuing post-war problems with printing, strikes, paper shortages, irregular publication dates and postal delays and losses. Four issues per year were maintained but the publication dates were changed from March, May, July and November to January, April, July and November to fit the school year better. One important and useful change introduced by the new editor was the identification of the author of each article: unfortunately this information is not given in the early issues of the Association's journals and the archival copies need annotating before the knowledge is lost.

The post-war change in the content already noted was maintained with a new balance between home and overseas regional articles and systematic and teaching articles. Physical geography however is still dominated by climatology and meteorology and only the occasional geomorphological contribution is found. One important innovation in 1954 was the introduction of a section on 'This Changing World' edited initially by L. S. Suggate: this consisted of short, one page accounts of changing geographic conditions world wide. This method of up-dating textbooks proved extremely helpful to teachers and 'This Changing World' has persisted as a feature of *Geography* to the present day.

*Geography* for this period contains some

notable and memorable articles. The March 1951 issue has E. W. Gilbert's 'Seven Lamps of Geography', an appreciation of the teaching of Sir Halford Mackinder which can still be read with profit by all intending geographers. The January 1953 issue includes papers on the Exmoor Storm of August 15 1952 by C. Kidson and Joyce Gifford; whilst the July 1953 journal has a definitive account of the North Sea storm floods of 1st February 1953 compiled by several authors. In a different vein, but undoubtedly one of the classic model accounts in *Geography*, we find in the April 1954 issue D. L. Linton's 'Landforms of Lincolnshire'.

Sadly the Geographical Association was not without losses amongst its senior ranks during this formative stage: deaths included Dr. Hugh Robert Mill in 1950 after a close connection with the Association from its foundation in 1893; J. L. Holland in 1952, first elected to the Association Committee in 1903 and subsequently important in promoting the concept of land utilisation survey; James Fairgrieve in 1953, whose 47 years of devoted service was quite exceptional; Dr. O. J. R. Howarth in 1954 was also closely connected with the Association for over 40 years; and finally Sir John Linton Myres in 1954, who although a historian had many links with geography (see Appendix K).

The loss of Sir William Himbury (Treasurer), Professor Sir John L. Myres (Trustee) and Lord Rennell (Trustee) in 1954–55 necessitated replacements and Professor L. Dudley Stamp (Treasurer), Professor E. G. Bowen (Trustee) and Professor W. G. V. Balchin (Trustee) were recruited to fill the vacancies.

The post-war expansion of geography within the universities was further consolidated by the creation or replacement of chairs in nine more universities between 1951 and 1954 so that by the mid 1950s Departments of Geography existed in nearly all the then universities in the United Kingdom (see Appendix L). The next phase however was to witness problems as the new universities began to emerge.

By now the United Kingdom was beginning to show signs of real recovery from the devastation of World War 2: the rationing of food, clothes and petrol had ceased and the recon-

Photo 19. Professor Emrys George Bowen 1900–1984. Trustee 1955–1973; President 1962.

struction of the bombed cities was well under way: apart from difficulties over foreign travel because of currency shortages there was an increasing air of normality in the life of the nation. In the educational sector settled conditions began to lead to thoughts of improvement and expansion of existing facilities and this trend applied particularly to the universities.

The Geographical Association itself also entered upon a more stable phase after the dislocation produced by the change of officers and the move from Manchester to Sheffield: the pattern of activity which had emerged in the early fifties continued steadily through the rest of the decade responding to membership needs and the economic state of the country. The membership climbed steadily as the position of geography in the curriculum improved and numbers in the teaching profession grew in response to the post-war baby boom then moving through the educational system. From 3781 full members in 1955 there was a steady increase to 4109 in 1956, 4366 in 1957, 4610 in 1958, to 4919 in 1959, with the magic figure of 5000 being passed to 5110 in 1960.

The pattern of Annual Conferences at the London School of Economics in January combined with a rapidly expanding publishers' exhibition continued under the able direction of R. C. Honeybone and Dr. J. H. Bird; Spring Conferences were held at York (1955), Brighton (1956), Matlock (1957), Aberystwyth (1958), Leicester (1959) and Durham (1960). Summer Schools were organised in Aix-en-Provence (1955 and 1957), Exeter and Guernsey (1956), the Scottish Highlands and Middle Rhineland (1958) followed by studies in Alpine geography in Switzerland and Mediterranean geography in Spain (1959). Travel restrictions on individuals because of currency problems made the overseas ventures popular as specially favourable financial arrangements applied to business and educational travel. Responsibility for the organisation of the Spring Conferences normally fell on the local branch where such existed, whilst the Summer Schools were undertaken by volunteer specialists from the universities. By the end of the decade attendance at the Annual Conferences was approaching 1,000, Spring Conferences attracted an average of 200 and the Summer Schools a further 200 participants. The Geographical Association was clearly providing valuable educational experience and facilities for its members.

The number of Branches also continued to increase steadily, reaching a total of 56 in 1959, partly as a result of efforts to increase coverage which flowed from a conference of 42 Branch representatives in 1955. Successful efforts were also made to resuscitate lapsed branches and to cross fertilise branches with ideas on lecture/field programmes and other activities. A number of overseas geography teacher associations also sought affiliation at this time, notably in Kenya, Sierra Leone and Jamaica.

The Presidents for 1955 and 1956 were Mr. L. S. Suggate and Lord Nathan but thereafter the post reverted to the professoriate, being filled by Professors P. W. Bryan (1957), R. Ogilvie Buchanan (1958), J. A. Steers (1959),

and A. Austin Miller (1960). With the death of Sir William Himbury in November 1955 the Association lost its long standing Honorary Treasurer—this appointment dating back to 1931: the Treasurer's duties were thereupon taken up by Professor L. Dudley Stamp, one of the Trustees. Another casualty in 1955 was Mr. T. C. Warrington, then in his 86th year, who was forced to retire completely owing to ill health; his place as Honorary Librarian was taken by Mr. L. J. Jay. Remarkably Professor Fleure, still active as Chairman of Council, reached his 80th birthday and completed 40 years of continuous service to the Association in 1957. A birthday subscription list was opened by the Association and sufficient funds raised to underwrite a visit by Professor and Mrs. Fleure to Norway. Accompanying the cheque was a book of signatures from over 300 subscribers. Professor Emeritus E. G. R. Taylor also attained the age of 80 in 1959 and the Association assisted in raising funds to endow an annual lecture in appreciation of her services to geography. In the same year Professor S. W. Wooldridge became a Fellow of the Royal Society and Professor L. Dudley Stamp received an Honorary Doctorate from the Stockholm School of Economics, Sweden.

The Section Committees, by now established, continued the programmes initiated in the early fifties: the Public and Preparatory Committee's main concern was with the Common Entrance examination paper and the organisation of conferences in Oxford and Cambridge; the Primary Schools Committee was active with the revision of the Primary Handbook which eventually appeared in 1959 with a new title *Teaching Geography in Junior Schools*; the Secondary Schools Committee was the most active, dealing with problems of untrained teachers in the Secondary Modern schools, the place of geography in the Technical schools, problems related to field work, GCE Ordinary level and GCE Advanced level syllabuses, the publication of a geography room survey leading to the supply of information to the Ministry of Education for their building bulletin, and finally a memorandum to the Joint Matriculation Board on Geography.

The Further Education Committee collected data on the state of Geography in Further

Education, co-operated with the Royal Geographical Society in making representations to the Associated Examining Board on their GCE examinations, examined the place of geography in Technology, and finally produced a Bulletin; the Training College Committee concentrated on organising three one day training conferences each year covering varied topics such as content, syllabuses, the place of radio and TV and aerial photographs in training, together with the planning and equipping of a geography room. Urban Spheres expanded its survey work from England into Scotland; Visual Aids began to explore the use of Ordnance Survey map extracts with air photographs and organised displays at the Annual Conferences; Field Studies, formed in 1955, concentrated initially on the organisation of summer schools and the collection of data about field study centres.

The Executive Committee had to cope with the problems thrown up by the Section Committees and additionally concerned itself with the emerging Association of Commonwealth Geographers, negotiation with LEAs for the recognition of geography courses for financial assistance, a submission to the Association of Education Committees deploring the low level of allocations for books and maps, a memorandum to the Central Advisory Council of the Ministry of Education on the place of geography in the education of the 15–18 year old age group, and finally the details of the proposed new series of booklets dealing with the landscape in selected areas in relation to the One Inch Ordnance Survey maps. Eventually entitled 'British Landscapes through Maps' this new venture was edited by Professor K. C. Edwards and proved very popular with teachers. The first publication on *The English Lake District* by Professor F. J. Monkhouse appeared in 1960 and over the next eighteen years nineteen such booklets were produced at a rate of roughly one a year: these covered the classic landscape areas of Great Britain. This series would have been of considerable interest to Dr. H. R. Mill who had advocated the notion of sheet memoirs for the One Inch OS maps at the beginning of the century.

During this period the Geographical Association was also involved through its indi-

Photo 20. Professor Kenneth Charles Edwards CBE, 1904–1982. Editor, British Landscapes through Maps, 1960–1978; President 1963.

vidual members with two national surveys. In January 1958 Professor David Linton published plans for a morphological survey of Great Britain leading to the production of morphological maps. This project was eventually taken up by the geomorphological fraternity. Of greater interest to teachers of geography was the second national project for a new Land Use Survey launched by Miss Alice Coleman at the Annual Conference in January 1960. The Second Land Utilisation Survey was a much more ambitious project than the First as it involved over 60 categories of use instead of the basic 8 of the First, and it envisaged publication of the results on a scale of 1:25,000 (a scale not available for the First Survey of the 1930s). With the enthusiastic help of individual teachers most of England and Wales was resurveyed in the early 1960s on the 1:10,000 scale producing a valuable data bank of information from which over 120 1:25,000 maps were subsequently published with the aid of grants from Local Education Authorities and charitable organisations.

There is an interesting increase in the number of articles in *Geography* devoted to land use both at home and overseas in the period which culminated in the launch of the Second Survey: as well as those articles directly concerned with land use there was also a marked increase in agricultural topics.

*Geography* had been given a face lift in 1955 with a new red cover but the Mollweide projection logo was still retained. The most important change however was the introduction in 1954 of 'This Changing World' which presented in each Journal five or six short up-dates of relevant topics from all over the world. The first editor of this section was Mr. L. S. Suggate who reluctantly had to resign from this post in November 1956 as a result of a family migration to New Zealand. From January 1957 the task was undertaken by G. J. Butland. With the addition of home and overseas articles as well as a steady stream of contributions related to aspects of teaching, plus the Association news, *Geography* had by now become an essential part of the geography teacher's armoury. The variety and range of the contents taken over a period of years is truly immense.

There are several seminal contributions of note at this time. Professor Stamp undertook a timely review of 'Major Natural Regions: Herbertson after Fifty Years' in his Herbertson Memorial lecture of 1957 (*Geography*, No. 198, Vol. XLII, Pt. 4, November 1957), whilst Professor E. W. Gilbert has a memorable article on 'The idea of the Region' in the July issue of 1960 (No. 208, Vol. XLV, Pt. 3). In a quite different vein there is a moving tribute to Professor H. J. Fleure by Professor Bowen in *Geography* for July 1957 on the occasion of the 80th birthday celebration.

The production of the Journal on time was not however without its problems. Throughout 1959 and the early part of 1960 the printing industry was bedevilled by strikes and issues came out at irregular intervals or as combined issues. There was no loss of content but editorial life was something of a nightmare on all journals during this phase.

Happily, apart from Sir William Himbury, the Association suffered no losses at the senior

level during this period. The retired professoriate however had several casualties, notably Professors Rudmose Brown (1957), Herbert King (1958) and W. W. Jervis (1959). There were also five incoming professorial changes as a result of retirement and migration during 1957–59. (See Appendix L.)

Towards the end of the decade the welcome increase in membership already noted brought with it a new problem as it became apparent that the accommodation at Duke Street was too small for the increased volume of work being processed by the staff. By this time over 40,000 items per year were being despatched to members of the Association from what had become very cramped conditions. Temporary relief was obtained in 1959 by a major reorganisation of the layout of the office and library aided by a magnificent gift of oak shelving from Sheffield University which solved the book storage problem. However the long period of some 67 years in which the Association had depended upon charity accommodation from educational authorities was about to end, as the need for increased space coincided with a warning notice from the City Fathers in Sheffield that the upper floor of the Duke Street Library would soon be required for other purposes. The moment had come when the Association would have to look for a more permanent home of its own and this would inevitably mean an increased annual subscription from its members.

## The Move to Fulwood Road

The early years of the 1960s were anxious ones for the Geographical Association's officers as the various options regarding accommodation were examined and in the main rejected—the leap to a really professional organisation owning its own property was then considered too big an undertaking but a solution had to be found. There was a temporary respite with the Duke Street site when the two years warning expired but a formal notice was received from the Town Clerk of Sheffield in September 1963, in the 70th year of the Association's existence, that the 'tenancy' of the Park Branch Library would have to terminate in September 1964. By late 1963 however Professor Alice Garnett had been able to enlist the aid of the

University of Sheffield. In acquiring land for halls of residence the University had also taken in a substantial Victorian house at 343 Fulwood Road which it was prepared to let to the Association at a reasonable rent. This solved the problem and the move took place during the summer of 1964 after 13 years of rent free accommodation in the Park Branch Library—for which the Association was and remains greatly indebted to the Sheffield City Council.

The formal opening of the new Headquarters at 343 Fulwood Road took place on 11th December 1965 by the Lord Mayor of Sheffield followed by a lunch given for University Officers to show them how the building was being used. There was thus a happy coincidence with the Herbertson Centenary Year celebrations. These had just concluded the previous day in Edinburgh where a joint meeting of the Royal Society of Edinburgh, the Royal Scottish Geographical Society and the Geographical Association had gathered to hear a commemorative lecture by Professor F. Kenneth Hare.

The accommodation problem occurred along with a period of intense activity in the Geographical Association's affairs. The membership continued its steady rise from 5110 in 1960 to 5349 in 1961, 5391 in 1962, 6177 in 1963, 6704 in 1964, 7357 in 1965, 7883 in 1966 and 8448 in 1967, all of which increased the administrative load at Headquarters considerably. In the same vein annual income/expenditure grew from £9380 in 1960 to £21,833 in 1965 with assets rising from £14,164 in 1960 to £27,994 in 1965. The annual subscription had been raised to £2 as from September 1961 in anticipation of the need for a new Headquarters location. The financial figures for 1966 are distorted owing to the inclusion of the very large sums arising from the Puerto Rico summer school of that year (see below).

Additionally the Secretariat had to service the Council, Executive Committee and numerous Section Committees, Branch business and much unreported work concerned with the promotion of the subject. The most notable activity in the last category was the 'behind the scenes' lobbying during the emergence of the new universities.

The University College of North Staffordshire, later the University of Keele, had begun life along conventional lines after its opening in 1950 (the foundation was in 1949): it included a Department and Chair of Geography initially filled by S. H. Beaver. The University of Sussex, founded in 1958, ushered in however a new era of university academic structures with inter-disciplinary approaches (Asa Briggs' 'New Map of Learning'). Geography had a part to play but not as a separate subject based on a separate department. In April 1959 the University Grants Committee set up a New Universities Sub-Committee and during the next three years capital grants were approved for additional universities at York, Norwich, Colchester, Canterbury, Coventry and Lancaster. In each case the promotion committee gave way to an Academic Planning Committee and each without exception followed the Sussex model of trying to provide a different academic structure from the conventional departmental and faculty approach.

Despite lengthy representations from outside bodies and more especially the Geographical Association, geography fared badly in the initial stages as it never secured a place on any of the Planning Committees. This was somewhat paradoxical as the subject had by now become firmly established in all the older universities and nationally was riding on the crest of a wave of popularity with the growth of interest in the environment and the publicity surrounding the 20th Congress of the International Geographical Union held in London in 1964. Eventually its importance was realised in the seven new universities and service appointments were made which have led in some cases to Departments or Schools of Geography or Environmental Studies being created.

During this period two of the four yearly meetings of the International Geographical Union occurred near enough to produce an impact on the Geographical Association. In 1960 the 19th IGU Congress was held in Stockholm and as already noted in 1964 the 20th IGU Congress came to London. It was thought inadvisable to attempt any Summer Schools in either of these years as so many senior geographers were involved in the IGU. Furthermore whilst the 1964 Congress in London was in process the Association had to

Photo 21. 343 Fulwood Road, Sheffield, the present Headquarters of the Association in the grounds of the University of Sheffield's Halls of Residence complex.
Photo: Geographical Association archives.

cope with the domestic move across Sheffield additional to its contribution to the London celebrations.

The professorial pattern in the presidencies was largely maintained during the early 1960s although Geoffrey E. Hutchings occupied the post in 1961 after Professor Austin Miller's tenure in 1960. Professor E. G. Bowen officiated for 1962, Professor K. C. Edwards for 1963 and Professor D. L. Linton for 1964—the critical year of the Sheffield move and the London IGU. Thereafter senior Association members

reappear as J. A. Morris was President in 1965, Professor S. H Beaver in 1966 and H.M. Inspector E. C. Marchant in 1967. (See Appendices C and K.)

The established conference pattern was also continued with the Annual Conference being held each year in London at the London School of Economics but with a change of organisers in 1964 as R. C. Honeybone was appointed to a Chair of Education in the University of Dar-es-Salaam in 1963. Dr. P. Odell and Dr. D. Brunsden then took over for

1964 and 1965, with Dr. Odell, J. A. Davies and C. R. Morley for 1966, and then back to Dr. Odell and Dr. Brunsden for 1967. Spring Conferences were held in Durham in 1960, Bristol 1961, Keele 1962, Swansea 1963, Falmouth 1964, Birmingham 1965, Oxford 1966 and Sheffield 1967. The 1964 Falmouth Spring Conference will be long remembered for the nearly disastrous expedition in the old *SS Scillonian* to the Scilly Isles—few of the 200 participants survived intact one of the roughest passages for many years!

There was also an Easter School in Italy in 1961 and Summer Schools were organised in the Ruhr and Black Forest in 1961, Norway in 1962, Scotland in 1963, Malta in 1965 and Puerto Rico in 1966. Although there was no formal summer school in 1964 owing to the IGU Congress in London members of the Association were not left without travel opportunities. The Isle of Thanet Branch under the Presidency of Miss Alice Coleman and the Secretary Miss Marjorie Woodward had for some time been thinking of the possibility of visiting North America using charter aircraft and charter buses. These ideas matured in the early 1960s and arrangements were made for such a visit in August 1964. The study tour was then opened to the whole Association and the hundred places were rapidly filled. The three week tour took in Montreal, Ottawa, New York, the Appalachians, Washington, Baltimore and Puerto Rico in the West Indies. Thus began a memorable series of inter-continental study tours over the next eight years in which visits were made to Eastern North America (1964), the West Indies (1966), East Africa (1968), Trans-Canada (1970) and the Western Cordillera of North America (1972), involving chartered aircraft, chartered trains and chartered coaches to cope with up to 150 participants on each expedition. Two of the enterprises became official Association Summer Schools (the West Indies, 1966 and Trans-Canada, 1970). Two field study leaders (Professor W. G. V. Balchin and Miss Alice Coleman) took part in all five tours. The group travel, charter transport and charter accommodation, produced some remarkable low cost travel bargains so that these ventures were much sought after at the time. They only ceased when the travel industry began to offer

Photo 22. Geoffrey E. Hutchings (President, 1961) demonstrating field sketching on a Countryside Course at Preston Montford Field Centre in 1963.

similar packages and when group travel became possible on scheduled flights.

The Isle of Thanet Branch was not alone in 'spectaculars' over this period. Liverpool had long claimed a branch membership of over 900 with record lecture attendances of over 500 and regular charter steamers on the Mersey, but in 1964 Birmingham reported a branch membership of over 1,000. Bolton and District Branch introduced VIth Form Conferences in 1961 and this idea quickly spread to other branches and continues to the present day. In 1966 the Bolton Branch organised two VIth Form Conferences which were attended by 1,300 sixth formers from 50 schools. Branches also began publishing books of local interest—Coventry, Brighton and again the Isle of Thanet were in the forefront of this activity. In a different vein the Lincoln Branch organised the placement of a plaque in the entrance hall of the Queen Elizabeth Grammar School in Gainsborough on 4th May 1961 to commemorate the centenary of the birth of Sir Halford

Mackinder in Gainsborough on 18th February 1961. The total number of branches by this time was fluctuating around 70—up to four or five new or re-established branches appearing each year to counter-balance temporarily suspended units.

The Geographical Association's Standing Committees were similarly active. The Common Entrance examination continued to exercise the minds of the Public and Preparatory Schools section along with the preparation of a journal and the organisation of conferences at Oxford and Liverpool. Two prizes, the Steers and the Lindsay, were also on offer for the 500 schools registered with this section. Primary Schools section coped with the problems of school journeys, co-operation with the BBC for a series on 'Exploration Earth', geography in infant schools, map work and the production of newsletters. The Secondary Schools section had a very heavy agenda: field study grants, syllabuses, memoranda for NUJMB and the London O level, books of exercises on OS maps, geography rooms, UNESCO, Schools Council, the use of atlases, Advanced level syllabuses, local studies, the raising of the school leaving age, and the new CSE examination were some of the problems dealt with. The Committee also lost in 1962, after 35 years of continuous service, Mr. C. B. Thurston, who was its first Chairman from 1927–1952. The Training College Section maintained a steady programme of three day conferences each year, of which two were normally in London and one in the provinces at locations such as Salisbury (1960), Scarborough (1961), Cheltenham (1962), Matlock (1963), Liverpool (1964) and Southampton (1966). Further Education also organised conferences in Salisbury (1960), Shrewsbury (1961), Castleton (1963), Luton (1964), Southampton (1964) and Luton again in 1966: additionally a newsletter was produced and close relationships established with H.M. Inspectorate.

Teaching Aids and Field Studies combined for discussions on the need for a Field Studies Manual but progress was brought to a halt with the sudden and successive deaths of two chairmen of Field Studies—Professor S. W. Wooldridge in 1963 and Geoffrey E. Hutchings in 1964. The Field Studies committee was reconstituted in 1965 under Dr. Ted Yates' chairmanship.

Several of the Section Committees and also the Executive were further engaged in the preparation of memoranda for submission to the Plowden committee on Primary Education which was then sitting. A new problem also emerged in the overlap between VIth Form geography and First Year University geography which led to the emergence of a new standing committee to deal with this difficulty. The notice to vacate the Duke Street premises precipitated the raising of the annual subscription to £2 as from September 1961 and this led the Executive Committee to reconsider the Statutes and Standing Orders: the modifications were adopted in January 1962 and the Geographical Association thereupon registered under the Charities Act of 1960. A memorandum was also prepared for submission to the Central Advisory Council for Education of the Ministry of Education on the role of geography in the education of children of average and less than average ability. The Association was also pleased to receive an invitation from the Ordnance Survey to send a delegation to the Map Users' Conference which began in 1962.

One way and another there was a certain amount of euphoria in the geographical profession and in the Geographical Association with the progress of events in the early 1960s. There was a big demand for the increasing number of the Association's publications, more especially reprints of *Sample Studies*, the new *Teaching Geography in Junior Schools* and the revised *Geography in Secondary Schools* Handbooks, along with the new 'British Landscapes through Maps' series. There was also co-operation with the Ordnance Survey in the production and sale of water and contour only pulls of certain maps. But the Association was about to undergo a further series of shocks with staff changes in the mid 1960s. Miss Marguerita Oughton resigned for personal reasons in November 1964, with effect from February 1965, after 14 years service as Assistant Secretary and Assistant Editor. There is little doubt that Miss Marguerita Oughton had been a king-pin in the organisation of the GA during a period of considerable growth as is clear from Professor Linton's remarkable five page tribute to her in *Geography* for April 1965. The replacement was Mr. Alec C. Smith who was appointed as Administrative Secretary.

In April 1965 Professor David Linton resigned after 19 years of service as Editor: his replacement was Professor Norman Pye of Leicester who had long served on the Editorial panel and whose connection with the Association went back to pre-war days when he was on Professor Fleure's staff in Manchester. A third blow came in the following year when Professor Alice Garnett intimated that she would be leaving her post of Honorary Secretary on her retirement from University work in 1967.

The concluding years of Professor Linton's editorship of *Geography* saw a consolidation of the changes which he had gradually introduced from 1950 onwards: articles on physical geography, especially geomorphology, climatology and biogeography became much more frequent and with the easing of the paper restrictions the very popular section on 'This Changing World' grew from three or four contributions in each issue to seven or eight: the responsibility for this section passed to Miss Oughton in November 1959 when Dr. G. J. Butland left for Australia. Subject matter for the articles and 'This Changing World' was world wide and covered the complete spectrum of geographical interests. Additionally each issue of the journal contained features on field work, teaching problems, apparatus, etc. as well as the much appreciated reviews of new publications. This period also saw the introduction of special theme issues, the most notable being the special publication of July 1964 to mark the 20th IGU Congress in London, and that of November 1965 to mark the centenary of the birth of A. J. Herbertson. Other seminal articles include Professor Beaver's 'History of the Le Play Society' in the July 1962 Journal whilst the annual Presidential addresses usually provided the basis for a group of related articles in at least one issue each year.

The Geographical Association faced and overcame two major external problems in the mid 1960s, both of which occasioned much heart searching at the time. On the one hand Branch activities were challenged in some areas by the appearance of geography groups within the newly formed County Teachers Associations with their regional teacher centres. This official move to promote greater co-operation generally was introduced with good intent but cut directly across some existing organisations of which the Geographical Association was one. The eventual solution before the scheme faded was to arrange for various forms of affiliation and co-operation with the Association.

Of greater danger to the Geographical Association however was the impact of the 'Quantitative Revolution' which by this time was beginning to affect the teaching of geography in the schools. Quantitative techniques were not by any means new (the need had been urged by some geographers before World War 1) but the discovery of statistics and the application to geographical problems by some of the younger geographers in the 1950s, combined with computer technology, opened up new possibilities of performing onerous statistical work at high speed. Sensing another 'new geography' quantitative techniques were enthusiastically taken up by many younger teachers but were viewed critically by the 'establishment' which was less familiar with the necessary mathematics, statistics and computer technology. At first the Association did not respond to the movement but eventually, realising that a gulf was developing within the profession, a new committee was formed in 1967 on the 'Role of Models and Quantitative Techniques in Geographical Teaching' under the chairmanship of Professor S. Gregory. This move began to bring back the 'revolutionaries' into the fold as the Committee was given wide terms of reference in an effort to resolve what had become a somewhat heated debate. A full report of the initial work of the Committee will be found in the January 1969 issue of *Geography*—which is almost entirely devoted to the problem.

On a happier note Professor Fleure, still acting as Chairman of Council, attained his 85th birthday on 6 June 1962 and was presented with a book of congratulatory messages and good wishes from several hundred Association members, together with a portrait in oils which now hangs at Headquarters. Professor Dudley Stamp was awarded an Honorary degree of D.Sc. from the University of Warsaw in 1962 and was knighted in January 1965 whilst

President of the Royal Geographical Society.

It is not surprising that with the progress of time the Association continued to lose a number of its senior citizens during this period. The sudden death early in 1960 of Mr. G. J. Cons robbed the Association of its President-elect for 1961: Mr. Cons had for many years played an active role in the work of the Association especially in relation to visual aids. His replacement for 1961 was Mr. Geoffrey E. Hutchings who was elected in recognition of his outstanding work for Field Studies: sadly he died in 1964. Other losses included Professor S. W. Wooldridge (President 1954) who died aged 62 on 25th April 1963, and the almost indestructible Mr. T. C. Warrington (President 1942–45) who passed away on 26th October 1963 at the age of 93, and Sir. E. John Russell (President 1923) who died 12th July 1965 also aged 93 (he had been leading study tours abroad up to the age of 86!). These were quickly followed by Professor F. Debenham (President 1952) on 24th November 1965 aged 82, and Professor J. F. Unstead on 28th November 1965, aged 90. Professor Unstead's passing revived memories of a very early link with the Association as he was Correspondence Secretary and assistant to Professor A. J. Herbertson from 1906–1915.

The most unfortunate loss for the Association came on 8th August 1966 with the unexpected death of Emeritus Professor Sir Dudley Stamp at the early age of 68 whilst attending an International Geographical Union Regional Conference in Mexico City. Sir Dudley (President 1950) was the Association's Honorary Treasurer and one of its Trustees and had been closely connected with its activities for nearly forty years. With a flat in London, a main residence in Bude, Cornwall, and a country cottage in British Columbia, Sir Dudley was a 'global' geographer in both outlook and practical ways. Always on the move, there was one memorable occasion when he 'dropped in' to an Executive meeting at the London School of Economics between changing planes at Heathrow on his way from Stockholm to New York. With his international as well as national contacts at the highest levels there is no doubt that his death was a major loss to British geography as well as the Geographical Association. In due course the Association issued an index to Volumes 1–54 of *Geography* as a permanent memorial to all his services for British geography.

Professor M. J. Wise was the replacement as Treasurer within the Association. Other notable deaths in 1966 included Sir Alexander Carr-Saunders (President 1947), Professor E. G. R. Taylor and Professor A. C. Odell.

Although developments in the new universities were not wholly favourable so far as geography was concerned the need to staff the new institutions broke the tradition that professors were committed to the university of their first appointment for their lifetime. Professors suddenly became mobile. Expansion of the numbers in the established universities also encouraged the creation of second and personal chairs. Combined with normal retirements we suddenly find a considerable movement taking place along with a spate of promotions. Between 1961 and 1967 no less than 30 new Professors of Geography appeared in the universities as a result of these changes (see Appendix L for details).

Coming shortly after the retirement of Professor David Linton from the Honorary Editorship the further retirement of Professor Alice Garnett from the Honorary Secretaryship at the end of 1967 marks the next distinct watershed in the history of the Association. Up to 1947 the two posts had been combined and held by Professor Fleure for 30 years. The post-war growth in membership and activities clearly pointed to a division of the duties and for the greater part of the next 21 years these were shared by Professor Garnett and Professor Linton. Many problems had to be overcome, not least that of accommodation and it is to Professor Garnett that the Association owes the greatest debt of gratitude for her successful negotiations over the Park Branch Library and Fulwood Road premises. Another major problem which was faced and overcome was the necessary revision of subscription rates as a result of the post-war inflation. Although now accepted as a regrettable fact of life inflation was then a new financial problem not readily assimilated: the change was eased however when negotiations with the Inland Revenue allowed subscriptions to be deducted from taxable income.

Membership during the Garnett-Linton

Photo 23. An interesting clutch of geographic luminaries captured on film on the occasion of the luncheon party at The Dorchester Hotel given for Mr. E. O. Giffard's retirement in 1965. From left to right: Mr. George Philip, Chairman of the George Philip Group of Companies; Dr. Edward Hindle, Honorary Secretary of the Royal Geographical Society; Mrs. E. O. Giffard; Mr. E. O. Giffard, Educational Adviser to George Philip & Son Ltd.; Professor Sir Dudley L. Stamp, President of the Royal Geographical Society, Treasurer and former President of the Association; Mrs. Audrey N. Clark, Editor, Longman Publishers, Dr. E. C. Marchant, Chief HMI Geography, later President of the Association in 1967; Mrs. George Philip; Mr. J. N. L. Baker, Former Reader in Geography at Oxford University and then Deputy Lord Mayor of Oxford.
Photo: Geographical Association archives.

phase consequently doubled from around 4,000 to over 8,000. Professor Garnett also strengthened and encouraged the work of the Section and Standing Committees and was a strong advocate of the benefits which the Annual Conferences, Spring Conferences and Summer Schools provided for the members: she also encouraged the work of the Branches. When Professor Linton left Sheffield for Birmingham in 1958 Dr. Garnett became Head of the Sheffield Department of Geography and had this further responsibility as well as that of the Honorary Secretaryship. The duopoly also witnessed a period of growth in the publications sponsored by the Association. *Geography* appeared in a new and enlarged format, 'British Landscapes through Maps', 'Sample Studies', 'Teaching aids' and numerous information pamphlets provided essential data for members. To both of these Honorary Officers the Association undoubtedly owes a great debt of gratitude for all their work over a long and somewhat arduous period.

# The Sixth Phase

## The Last Quarter Century 1968–1993

The last quarter century of the Geographical Association's history is marked by a further change in organisation. Whereas in the early days membership was small and much depended on voluntary service from a few individuals over lengthy periods the Association entered a time in which membership grew to a substantial number, where the potential for voluntary service became much greater and where subscriptions were sufficient to employ adequate supporting administrative and secretarial staff. With Professor Garnett's retirement and as from the beginning of 1968 the responsibilities of the Honorary Secretary's post were shared between two officers—initially W. R. A. Ellis and S. Gregory: happily Professor Garnett was persuaded to accept the Presidency for 1968 in order to ease the change over. The concept of a fixed term of office rather than an indefinite appointment also began to emerge at about this time: this was later to have a considerable effect with the next revision of the constitution.

Another reminder of the passing of the old order came with the death of Professor Fleure on 1st July 1969 at the age of 92. His decease severed the last remaining link with the formative years of the Association. He had remained active throughout his 'retirement' and was still Chairman of Council at his death. He was writing and publishing in his 90s although he once said to the present writer that he was then mainly engaged in preparing obituaries for the next generation of geographers—having already exhausted the list of his own contemporaries! Proof of this can be seen in *Geography* for November 1969 where tributes to Professor Fleure from Professors Bowen and Garnett are followed by an obituary notice for John Kirkland Wright signed by H. J. Fleure.

There is no doubt that Professor Fleure will rank as one of the great British geographers of the twentieth century and the Association was fortunate that James Fairgrieve and H. J. Mackinder between them persuaded him to

become Honorary Secretary in 1917. Professor Fleure's 63 years with the Geographical Association, of which 52 were in office with 30 years as Honorary Secretary and Honorary Editor, constitutes a record which must surely stand for all time. Very appropriately Professor Garnett became Chairman of Council with Professor Fleure's passing.

The new Honorary Officers very quickly found themselves wrestling with both new and old problems. Inflation continued to play havoc with all Society funding and the need for another relatively major rise in the subscription rate soon became apparent. The constitution was such that this could not be brought into operation until 1970 and the increase from the £2 fixed in 1961 to £3 had an immediate adverse effect on membership numbers which had not shown up before with much smaller increases. The membership had risen steadily through the sixties, reaching 8447 in 1967 and 8522 in 1968, to 9138 in 1969, but after the 50% increase in the subscription rate it fell back to 7991 in 1970.

Recovery came of course—membership was back to 8387 in 1971 and 8301 in 1972—but a new pattern emerged for future years with an immediate 10% loss when each inflation adjustment took place at intervals. Eventually an annual adjustment of small amounts was to prove necessary.

Another problem in June 1970 was the loss of the Administrative Secretary, Mr. Alec C. Smith, to Salford University as an Assistant Registrar after only five years' service; his departure was a reflection of the Association's increasing inability to match market rates of pay. His replacement from September 1970 was Mr. R. E. Fell who was then Registrar of Derby and District College of Technology: fortunately for the Association he wished to return to his native Sheffield.

All was not gloom however, in 1970 as this was the year of the memorable Trans-Canada Study tour in which 152 Association members spent a week in Montreal, followed by a week in Vancouver after traversing Canada in easy stages by Canadian National Railways. Charter flights were arranged from London to Montreal and Vancouver back to London. This was the largest and most ambitious overseas study tour ever organised by the Association and it produced a welcome and completely unexpected financial bonus for the Association as a result of a last minute change in the exchange rate (it could of course have gone the other way!). A necessary note of caution as regards future activities of this kind was soon forthcoming from the Honorary Treasurer, especially as the large sums of money involved in the chartering of aircraft, trains and coaches distorted the annual financial statement for 1970 (*Geography*, January 1971).

A wholly unexpected problem, not of the Geographical Association's making, arose in early 1971 with the lengthy national postal strike. This revealed the extent to which nearly all societies depended on the postal services for communication with their members. A major casualty for the Association was the Spring Conference in Southampton which had to be cancelled as the necessary information about numbers could not be collected nor the nature of the programme disseminated.

By this time the problems produced by the 'Models and quantitative Methods' revolution were being rapidly overcome as the newly constituted Standing Committee began to organise courses jointly with the Department of Education and Science at the Maria Grey College in Twickenham during the long vacations of 1969, 1970 and 1971. These courses, which were repeated several times each year to meet the demand, were on 'Statistical Methods and Models in Schools'. But a new problem was beginning to emerge in the late '60s as the environment became of public concern and 'environmental' courses began to appear in some universities and schools, together with related examinations. Syllabuses often revealed that in many cases the new courses were nothing more than geography under a new name, and furthermore were often being taught by geographers. As with the quantitative problem the Association responded by setting up a Standing Committee for Environmental

Studies in order to monitor the situation. Other problems which exercised the minds of the Executive at this time included the Metrication process and a revision of the Charter and Statutes to cover the new arrangements regarding the Honorary Officers, along with the revised annual subscriptions.

As already noted the President's post in 1968 was very appropriately filled by Professor Alice Garnett. She was followed in 1969 by Professor J. R. James, a geographer turned planner; then in 1970 by Mrs. I. M. Long, an active teacher, in 1971 by Professor W. G. V. Balchin, one of the Association's Trustees, and in 1972 by Alan D. Nicholls, another active Head teacher and also Assistant Honorary Treasurer.

The conference programme was considerably augmented at this time largely through the activities of the Standing Committees. The Annual Conference at the London School of Economics continued under the able direction of Drs. Odell and Brunsden in 1968, and then of Dr. Brunsden with D. K. C. Jones in 1969, 1970 and 1971. Spring Conferences were organised at Nottingham in 1968, Brighton in 1969 and Liverpool in 1970; but as already noted Southampton in 1971 was a casualty as a result of the postal strike and this conference had to be postponed to 1973, after Newcastle upon Tyne in 1972.

A varied programme of study tours, summer schools and field studies was available for members ranging from summer schools in the Benelux countries and the Middle East in 1968 to the Trans-Canada Study tour of 1970. Field studies courses were held at Slapton Ley and Rogate in 1968, Wymondham and Cwm Risca in 1970, Preston Montford and Charney Bassett in 1971, together with the already noted Statistical Methods and Models courses at Twickenham in 1969, 1970 and 1971.

The Geographical Association's Section Committees were equally active. The Public and Preparatory Committee organised Quantitative courses at Brathay Hall in 1969, continued the publication of *Notes and Queries* and examined the adequacy of certain O level syllabuses. Primary Schools Committee was concerned with the content of school atlases and produced a new Primary Schools Handbook. The Secondary Schools Committee considered school atlases, the place of Urban Geography in schools, revision of the JMB O level syllabus, a new Secondary Schools Handbook, and the impact of environmental studies. The Further Education Section continued its pioneer work with a series of conferences at Stevenage, Abergavenny (1968), London and Cambridge (1970) together with a consideration of links with the Open University, the Institute of Bankers, the Army School of Education, the BBC and ITV. The Colleges and Departments of Education section continued with their usual three annual conferences at centres including Nottingham, Hereford, London, Loughborough and Gloucester. The Teaching Aids Committee investigated the potential of the overhead projector and had discussions with the BBC about school radio. The Field Studies organised the summer study groups already noted above; the VIth Form Overlap committee produced reports (published in *Geography*) on the place of geomorphology, climatology and regional geography in the curricula at the VIth Form level; the Models and Quantitative Techniques committee continued its school survey and organised the courses at Twickenham already noted; whilst the new Standing Committee for Environmental Studies began its deliberations.

Branch life continued with a multitude of activities, both academic and social, although there were growing signs in some areas of a fall in adult attendance at the lecture programmes, which probably relate both to the spread of TV and to the demands on teachers' time in coping with school re-organisations and the proliferation of curricular and methodological developments. Excursions, field work, film evenings, VIth Form conferences, co-operation with other local organisations and social events proved popular however, with Branches tending to specialise in one or more activities. New or resuscitated branches appeared in Bradford, South-east Essex, Northampton, Canterbury, Dunfermline, Ealing, Londonderry, Taunton, Guildford, West Sussex and the Kingdom of Fife, more than balancing the temporary suspension of Carlisle and Cumberland, East Middlesex and Scarborough. The total number fluctuated around 75.

Although the balance and format of *Geography* inherited by the new Editor,

Professor Norman Pye, from Professor David Linton, left little scope for improvement three important changes were made: 1966 saw a big expansion in news items from Headquarters and the Section and Standing Committees, in 1969 all reviews were fully named for the first time, and in 1971 the width of the journal was increased to facilitate map presentation. 'This Changing World' continued to attract a steady stream of contributions, most issues containing four or five mini-articles in support of three or four major overseas contributions on a wide variety of topics and localities. There were perhaps fewer home articles but this was compensated for by many more methodological articles which reflect the discussion then in progress about the place of models, games, quantitative methods, statistics, the environment, and field work in the syllabuses. Seminal articles on the place of geomorphology, climatology, economic geography and regional geography in VIth Form work appeared in *Geography* between 1968 and 1971.

Having by now exceeded the proverbial three score years and ten in age it is not surprising that the Geographical Association continued to lose many of its senior members: losses in 1968 included Professor A. Austin Miller (President 1960) aged 67 and Professor P. W. Bryan (President 1957) aged 83: in 1969 as already noted Professor H. J. Fleure (President 1948) aged 92, C. B. Thurston (Honorary Member 1968), Professor R. H. Kinvig aged 75 and J. H. Stembridge aged 80: in 1970 L. S. Suggate (President 1955) aged 81, and most unexpectedly Professor D. L. Linton (President 1964) aged 64.

The already mentioned expansion and consolidation of geography at the professorial level in the universities continued and during the period from 1968 to 1972 another twenty-four new professors appeared (see Appendix L for details).

The 1970s were to witness fundamental changes in the organisation and work of the Geographical Association: some were forced by outside circumstances but most were of internal origin. The combination of these changes produced a major alteration in profile which has dominated the remaining years of the Association's first century.

The first major change came as a result of

Photo 24. A work scene at Headquarters. Professor Norman Pye (Honorary Editor, 1965–1979) checking proofs with Mrs. D. M. Ellis (Assistant Editor).
Photo: Geographical Association archives.

the Heath Government's decision to make January 1st a public holiday as from 1974. The long sustained tradition of holding the Annual Conference at the London School of Economics immediately after Christmas and over the New Year period became impossible as the University authorities closed the School on New Year's day: subsequently the closure was extended for a week or more as an energy-saving move. The inevitable Association committee set up to resolve this problem could only recommend that the Annual Conference be moved to immediately before or after Easter depending on school holiday dates. The change took place in 1976 (many non-members having turned up at LSE in January 1976 to find that there was no conference in session!). Sadly this meant a collapse of the Geographical Association's rhythm of an Annual Conference in the Christmas vacation, a Spring Conference in the Easter vacation and Summer Schools in the Summer vacation. No further Spring Conferences in the old tradition have been possible and Summer Schools have proved increasingly difficult to mount with rising costs and the many difficulties that have beset the teaching profession in recent years. The relative decline in salaries, the shortage of funds for vacation activities and the continual re-re-organisation in schools have all militated against voluntary self-improvement by the individual teacher.

The second significant change can be traced back to the establishment in September 1973 of an Organisation and Development Committee with the specific remit of reviewing the aims and functions of the Geographical Association and its Section and Standing Committees. This Committee produced a succession of reports which were duly considered by the Executive and all the Association's Committees over the next three years. By the middle of 1976 general agreement had been reached as to the format of a complete revision of the constitution, statutes and standing orders of the Association. The 'new look' was finally approved by the Annual General Meeting of 6th April 1977.

*The 1977 Constitution*
The new structure envisaged a governing Council with three standing committees covering Education, Publications and Communications, and Finance and General Purposes: supporting these were the Section Committees, Working Groups and Working Parties. The officers' posts were also redefined with two Honorary Secretaries and an Honorary Treasurer, one Honorary officer being attached to each of the three main standing committees. The President was also required to serve four years—two before and one after the Presidential year. A Publications Officer and Branch Officer were added in support of the Library and Information Officer, the Annual Conference Officer and the Trustees. New arrangements were also listed for the Editors of the Association's journals and publications.

Probably the most drastic change in the new Constitution was the agreement to limit the duration of tenure of the officers. The December 1968 Council and subsequent AGM had already approved minor constitutional changes specifying maximum periods of service for some officers but the 1977 revision placed a limit on *all* offices. Whilst this varied according to the office, generally a maximum of six years (two periods of three years) was imposed. This

brought to an abrupt end the tradition of individual life-long service to the Association which had characterised the first 84 years of its history: Henceforth long service would only be possible by changing office and even in these cases relatively short periods would result in comparison with the half centuries or more amassed by the Association's Founders. Whilst the change was hotly debated at the time (the compulsory retirement of effective and willing voluntary officers is always regrettable) it was perhaps a sign of the strength of the Association that it could now envisage a relatively rapid turn-over of officers to ensure a continued flow of new ideas and so prevent any incipient stagnation. By the early 1980s the last of the long serving officers (Professors Balchin and Pye) had been 'retired' since when the Association has been run essentially by its active teaching membership and the professoriate now has a much diminished role. The long term effect of this major change has yet to be assessed.

Another fundamental debate leading to change also took place in the early 1970s on publication policy. Although the Association was already producing *Teaching Geography* pamphlets in addition to the regular issue of *Geography* there was a call from teachers for more practical aids to teaching, material which could only be supplied by teachers. A plea for more teaching articles was inserted in the January 1973 issue of *Geography* by the Editor but this produced little response despite the initiation of a 'Teacher's Forum' section in the Journal from July 1973 onwards. Concurrently discussions were in progress with Longman on the production of a new journal to meet this need. Envisaged as a teacher orientated journal with Patrick Bailey as the Honorary Editor this project eventually got off the ground in April 1975 with the appearance of *Teaching Geography*. Initially a joint venture with Longman the new publication was in due course taken over entirely by the Association. With its appearance 'The Teacher's Forum' was dropped from *Geography* and the *Teaching Geography* pamphlet series gradually phased out. This was an interesting period for the few who remembered the great debate of the 1920s when *The Geographical Teacher* was transformed into *Geography*!

Although the mid 1970s thus saw fundamental changes in the life of the Geographical Association the early years continued the traditional rhythm of the post-war years: the professoriate supplied most of the Presidents with Professor Robert Steel in 1973, Professor Harry Thorpe in 1974, Professor Michael Wise in 1976–77 and Professor Stanley Gregory in 1977–78. The Presidency for 1975–76 was held by Miss Sheila Jones, a practicing teacher.

Annual Conferences continued at LSE in January under the direction of Dr. Brunsden and Mr. D. K. C. Jones from 1972–75 but changed to Easter in 1976. Spring Conferences were held at Southampton (1973), Worcester (1974) but Glasgow (1975) had to be cancelled owing to insufficient support (a reflection to some extent of the industrial problems of the mid 1970s). The planned Bradford 1976 Spring Conference had to become the Bradford 1976 Summer School as a result of the changed time of the Annual Conference.

The reports of Branch activities at this time reveal a very varied picture. Whilst some indicate a disturbing situation with regard to adult membership and attendance levels others reveal flourishing organisations. The Isle of Thanet staged in 1972 another spectacular study tour of the Canadian and American Rocky Mountains for 150 members in chartered aircraft and coaches: at least 13 branches mounted successful VIth Form Conferences, and the Birmingham branch achieved a record membership of 1,250: social activities were also very popular. The total number of branches continued to fluctuate slightly reflecting in most cases the voluntary enthusiasm of the local officers.

One important event occurred in 1970 which was to have subsequent repercussions at both branch and national levels: this was the formation of the Scottish Association of Geography Teachers by the 'Scottish Twenty'. The three main GA branches then in Scotland at Edinburgh, Glasgow and the Kingdom of Fife initially remained 'loyal' to the GA but eventually the reality of geographical distance, the different educational systems and the historical heritage strengthened the SAGT into becoming the voice for geography teachers in Scotland. But it was 1977 before the Glasgow and Kingdom of Fife branches, and 1979 before

the Edinburgh branch disappeared from the GA records. In the 1980s cordial relations were established between the GA and SAGT and there is now a considerable exchange of information of mutual interest whilst individual members are welcome at each other's conferences and SAGT participates in the Worldwise Quiz.

The membership figure fluctuated around the 8,000 mark varying from 8301 in 1972 to 8400 in 1973, 8522 in 1974 and 8470 in 1975, but dropping back to 7941 in 1976 and 7963 in 1977 as a result of further inflation related increases in the subscription rates in 1976. Although several hundred new members were joining the Geographical Association each year the inflow was being balanced by an outflow of retiring teachers and it began to look as if the Association had reached its optimum size for the time being.

Financially the annual income/expenditure also fluctuated around £48,000 for the years 1972–75 but jumps to £80,136 for 1976 as a result of the revised subscription rates combined with the appearance of the new *Teaching Geography* journal. Assets rose steadily from £66,619 in 1972 to £83,885 in 1976.

Despite the impending changes which loomed ahead in the discussions of the Organisation and Development Committee the 'old' standing and section committees continued with their work in the first half of the decade. We find concern with the environmental aspect running through many of their deliberations. The Environment at this time had suddenly become of public concern and 'environmental science' courses were appearing in schools, colleges and universities. The Primary and Middle Schools section were further interested in field work, the Third World, TV programmes and discussions with the Publishers' Council about the content of geography books for these levels. The Secondary Schools section agenda also included environmental studies, mixed ability teaching, insurance of school field groups, revision of A level syllabuses, problems of the proposed 16+, the O level and CSE examinations. The Public and Preparatory section continued discussions on the reform of the Common Entrance Examination, O levels and the 11–16 curriculum, and also held a conference in Oxford. The

Colleges and Departments of Education section maintained a vigorous programme of two or three conferences each year on a variety of topics. Further Education similarly favoured one day conference activity staging at least two each year in various localities. Field Studies had a mixed programme of committee discussion and field excursions and residential courses in conjunction with the Field Studies Council. A draft code of conduct was produced and the idea of a Field Studies Manual was mooted. Teaching Aid reviewed the commercial material available and surveyed the new Resource Centres. Models and Quantitative Techniques continued with the promotion of teachers' groups and the revision of bibliographical lists and investigated possible links with the Mathematical Association. The University/Sixth Form overlap committee turned its attention to social geography, eventually publishing a report in the April 1974 issue of *Geography*. Two new committees were established, the first to cover matters of interest in universities and polytechnics, and the second, environmental education.

As well as the flow of papers from the Standing and Section Committees the Executive was also concerned with the publication of the British Landscape Series, Teaching Geography pamphlets, the Dudley Stamp Memorial Index, A Bibliography of Geography in Education, proposals for the new *Teaching Geography* journal and the implementation of the new constitutional proposals.

*Geography* under the continued editorship of Professor Norman Pye maintained a steady presentation of informative articles on home and overseas geographical matters. 'This Changing World' continued its popularity with four or five concise contributions in each issue of the journal on a wide variety of topics. There was a noticeable increase in articles concerned with the environment whilst physical geography was more in evidence. Book reviews and Association News were also prominent.

As already noted the first issue of *Teaching Geography* with its distinctive space age cover appeared in April 1975 after a somewhat protracted gestation period. Initially five issues a year were planned but this was changed to four from August 1976 onwards. The annual subscription was fixed at £5. The pioneers who had fought the change of title of the Geographical Association's original *Geographical Teacher* to *Geography* in the 1920s would doubtless have expressed delight had they been able to see the appearance of *Teaching Geography* in 1975. Under the able editorship of Patrick Bailey of the School of Education in the University of Leicester a well illustrated new journal got off to a flying start and rapidly became essential reading for the geography teacher at the chalk face. Largely written by teachers for teachers some 12–15 articles in each issue were concerned with practical matters of teaching techniques, aids to teaching, field work problems and possibilities, syllabus discussion, resources, examinations, curriculum development, careers for geographers, book reviews and the teaching of geography overseas. Once teachers realised the possibilities of the new medium for the exchange of information there was no shortage of articles or material for publication. By Volume 3 July 1977 the pattern of four issues each year in July, November, January and April had been firmly established.

The constitutional changes introduced in 1968–69 began to take effect in the early 1970s. Professor Stanley Gregory relinquished the Honorary Secretary post after six years and was succeeded by Mr. G. M. Lewis in 1973. Professor Gregory continued as a Trustee however as Professor E. G. Bowen resigned at the end of 1972. Dr. N. J. Graves retired from the editorship of the Teaching Geography pamphlet series in 1973 with Mr. M. C. Naish taking over. Mr. L. J. Jay resigned as Librarian at the end of 1973 and was replaced by Dr. D. B. Grigg. Mr. A. D. Nicholls became Chairman of Council in place of Professor Alice Garnett and Dr. D. Brunsden retired as Honorary Conference Organiser in 1974. Mr. D. G. Mills replaced Professor M. Wise as Honorary Treasurer and Trustee and Mr. R. A. Daugherty succeeded Mr. W. R. A. Ellis as the second Honorary Secretary in 1976. Losses of senior members by death during this period were happily few in number and confined to Professor E. W. Gilbert (1973) and Professor F. J. Monkhouse (1975).

Continued expansion of geography within the universities was marked during the early

1970s by the appearance not only of second but also third chairs in both personal and established categories in a number of departments. Combined with replacements as a result of retirements there is a sudden surge of professorial appointments before the difficult retrenchment phase began in the universities. Between 1973 and 1976 no less than 28 new Professors of Geography were appointed (see Appendix L for details). These appointments all indicate the steady consolidation of the subject within the academic framework.

The combination of the changed time of the Annual Conference in 1976 and the new constitution in 1977 produced another definitive watershed in the Association's history. The annual rhythm of events noticeably altered with conferences and vacation schools being rescheduled and the new organisational structure producing a different pattern of meetings throughout the year. Responsibilities were also spread amongst a larger cohort of voluntary officers as the work of the Association expanded. After Professor S. Gregory (1977–78) the presidential office was filled by Dr. N. J. Graves (1978–79), Professor J. A. Patmore (1979–80), Mr. V. Dennison (1980–81) and Professor W. R. Mead (1981–82). The Honorary Secretaries over this period were Mr. G. M. Lewis (1973–79), Mr. R. A. Daugherty (1976–81), Mr. B. E. Coates (1979–84) and Mr. M. T. Williams (1981–85). These officers were supported by the newly created Publications Officer (Mr. R. A. Beddis), Branch Officer (Miss S. W. Jones), Library and Information Officer (Mr. T. W. Randle) as well as the Trustees, Honorary Treasurer, Annual Conference Organiser and the Editors of *Geography* and *Teaching Geography*. Professor W. G. V. Balchin was retired as a Trustee, a post he had held for 22 years, but was retained on Council because of his Ordnance Survey links. The Council had become the ruling body but the Executive had disappeared in favour of three Standing Committees covering Education, Publications and Communications, and Finance and General Purposes. The Chairmen of these three Standing Committees assumed increasingly active roles.

A number of important issues faced the newly constituted Council and its committees and Officers. Not least of these were the financial problems associated with the rapid rise in inflation at the end of the 1970s. The finances of the Association were becoming very susceptible to the changes in membership numbers; these dropped to 7941 in 1976 and 7963 in 1977 as a result of a further adjustment in subscriptions. Although there was a slight improvement to 8049 in 1978 and 8024 in 1979 there was a substantial fall to 7203 in 1980. With deficits of £5,569 in 1978 and £15,504 in 1979 it was clear that further subscription increases would be essential and that the Association was moving closer to the point where annual increases would have to be made to keep pace with inflation. Additionally the Association was also suffering from the cyclic nature of the expenditure/income factor arising from the publication policy which was tending to distort the annual accounts. Overall income/expenditure had reached £80,024 in 1977, rising to £111,731 in 1980 with assets at £91,678 in 1977 falling to £78,570 in 1980. Although the further subscription increases resulted, as noted, in a 10% membership fall to 7203 in 1980 there was a welcome increase from publication sales which took the Association's finances out of the red for that year.

Another time consuming problem which confronted the new organisation was the discussion within James Callaghan's 'Great Educational Debate' of the Q, N and F proposals of the School's Council's geography subject committee. Ever since the early 1960s the Schools Council had been considering ways of widening the 16–19 curriculum. Various proposals had been made to replace the A (Advanced) level examination in each subject with a Q (Qualifying) examination in the first year followed by an F (Further) examination in the second year, but these ideas were rejected in favour of a five subject combination at 18+ of which three would be taken at N (Normal) level and two at F (Further) level. The Association was asked to comment on the proposals and also provide specimen syllabuses: this led to a great deal of discussion in the appropriate committees and produced a number of critical articles in both *Geography* and *Teaching Geography* during 1977–79: these provide an interesting background to the national curriculum discussion which was to dominate the later 1980s.

The decade in fact began with the Education Standing Committee considering the response the Association should make to the Department of Education and Science proposals on *A Framework for the School Curriculum* and this led to the Association printing for general circulation 50,000 leaflets entitled *Geography in the School Curriculum 5–16*.

Another problem which was aggravated by the worsening national economic situation and the high inflation rate was the place of the Fleure Library in the Association's structure. Some 700–800 books were being added yearly to the already impressive collection and it was obvious that shelf space would soon run out. To complicate the matter postal library loans—once a mainstay of the Geographical Association—were falling away rapidly as postal rates began to rise steeply. By 1978 only 253 members were recorded as library users and most of these were personal callers from the Sheffield area. Although a library was seen to be an essential part of the Association's services by the Publications and Communications Standing Committee it was to be some time before a solution to the library problem was put forward.

One important external sector where the Geographical Association had successful dealings during this period was the Ordnance Survey. There has always been a close relationship between the two organisations as it is to their mutual interest to co-operate: the Association's members are to a large extent producing the Ordnance Survey's future market for maps. In the early 1960s the OS initiated an annual Map Users' Conference for consultative purposes: the Association became a prominent member sending several delegates normally led by Professor Balchin. In 1976 the Local Authorities Ordnance Survey Consultative Committee convened an advisory committee for the purpose of examining the requirements in Education for OS maps and services. Professor Balchin was also appointed to this committee and from it emerged not only an excellent Student Map Pack series but also a resolution of the vexed problem of the copying of OS maps. It was eventually agreed that Local Authorities would take out 'blanket licences' to cover OS copyright fees so that local authority schools would be free to copy for educational purposes without fear of copyright prosecution.

The Geographical Association was also a major contributor to the submissions prepared for the Government's Ordnance Survey review Committee (Serpell) appointed in January 1978 which reported in July 1979. One of the results of this review was a new OS Consultative Committee structure in which eight groups covered all the national interests. One of these was Education and Professor Balchin became Chairman of this Consultative Committee on which the Association also had a further representative along with all the other geographic and cartographic interests.

As well as these items of major concern the section committees of the newly formed standing committees dealt with a variety of matters. The Primary and Middle Schools had discussions with the Publishers' Council and the BBC on related topics and considered a new handbook for the 5–13 range: Secondary Schools reviewed the revised JMB A level examination, had discussions with the BBC and IBA about the geography content of TV programmes, and considered the supply of books and ethnic minority teaching. Higher Education was involved with environmental science, careers and establishing closer links with the National Conference on Geography in Higher Education: Further Education established contact with the Technician Education Council and the Business Education Council as well as organising several conferences: Teacher Education was heavily committed with the N and F proposals debate as well as with environmental studies and normal conferences. The new working groups covering Field Studies, Models and Quantitative Techniques, Environmental Education, and the Public and Preparatory Schools continued with their normal conferences and working activities.

The new Publications and Communications Standing Committee was established to develop the Association's publishing activities, to maintain library and information services, to approve the arrangements for conferences and to assist and promote branch activity. Some time was necessarily spent in its early existence mapping out a publication programme, an exercise complicated by the fact that a decision had to be made on the tradition that Association authors did not receive remuneration or royalties for work undertaken for the Association. After protracted discussion it was decided to maintain the tradition although it was realised that this would restrict the number of potential authors.

With *Geography* and *Teaching Geography* both well established attention was initially concentrated on new handbooks for teachers: a new Secondary School Handbook by Professor Norman Graves appeared in 1980 and a Primary and Middle School Handbook edited by David Mills followed in 1981. These were rapidly taken up by teachers thus providing a welcome boost to the annual balance sheet. *Geography into the 1980s* by Mrs. Eleanor Rawling was also a best seller. The already mentioned difficulty of the use of the Fleure Library continued as an unsolved problem and eventually in 1979 a working party was set up to try and find a solution.

One interesting development in 1978 was the establishment of limited term study groups working closely with the Association but funded by external agencies. The first of these was the Geographical Association Package Exchange Project (GAPE) funded by the Council for Educational Technology: this was followed by the 'Images of Canada' Project (ICP) funded by the Canadian High Commission.

Whilst the Geographical Association was thus making considerable progress on a number of fronts the enforced removal of the Annual Conference from the Christmas to the Easter vacation led to the collapse of the traditional spring conference—summer school pattern. From 1976 the Annual Conference continued at LSE at Easter under the able direction of D. K. C. Jones (1975–78) and Dr. B. S. Morgan (1979–82) but efforts to move the Spring Conference with its 200–300 attendance to the Summer months were not successful. A Summer Conference was tried in Bradford in July 1976 and a Summer Workshop in Leicester 1977 but the response was muted, and the Summer Schools planned for the Benelux countries and Aberdeen in 1978 had to be abandoned owing to lack of support. A further attempt was made in 1979 to hold a Summer Conference in Aberystwyth but this only attracted 24 participants. Smaller

section conferences, seminars and working parties continued in popularity but sadly the Heath Government had thrown a major spanner into the working of the Association with its decision to make 1st January a public holiday.

Branch reports throughout this period reveal continued financial problems because of inflation. Lecture programmes were still central to activities but increasingly difficult to mount in some areas as travel costs escalated: although all lecturers gave freely of their time they could not be expected to meet travelling expenses in addition. The exceptionally cold winter of 1978–79 did not help either. Branches were in many cases thrown back upon their own resources but this led to a certain amount of innovation with field work and social activities. The overall distribution remained generally static but with new branches appearing in Cardiff, North London, Central London, Ealing, Walsall and Northants balancing temporary closures in Portsmouth and more permanently in Edinburgh. The appointment of a Branch Officer in the Headquarters hierarchy and the appearance of a Branch Newspaper were important factors in holding the Branch fabric together and by the beginning of the 1980s more favourable reports from the Branches were being received at Headquarters.

The new Council which emerged in 1977 from the revised constitution resuscitated the concept of Honorary Membership for individuals with long and meritorious service. Before the change the records reveal the creation of only three honorary members by earlier Councils despite the considerable number of individuals who would undoubtedly have qualified by present day standards. Part of the explanation might well be that the Association could not afford this luxury in the early days. Dr. H. R. Mill had been made an Honorary Member in 1901, E. O. Giffard in 1965, and C. B. Thurston in 1968. The new Council began with W. R. A. Ellis in 1978 followed by Professor W. G. V. Balchin, Professor Alice Garnett and A. D. Nicholls in 1980 with subsequent additions (noted below) in most years in the 1980s.

*Geography* continued under the able editorship of Professor Norman Pye until his retirement at the end of 1979 by which time he had completed nearly 30 years of editorial service, having been originally appointed Honorary Assistant Editor from 1950–53 before becoming a member of the Editorial Panel (later restyled Board) from 1963–65, when he became Editor in succession to Professor D. L. Linton. He continued in service for another four years as an Honorary Vice-President: Dr. R. J. Small followed as Editor. With the appearance of *Teaching Geography* in 1975 *Geography* reverted to its traditional mix of general educational articles, home and overseas local and regional accounts, the ever popular short contributions in 'This Changing World' plus book reviews and news of the Association. The November 1978 journal however was a special Physical Geography issue reflecting the theme of the Annual Conference of that year. As in previous years the overall fare in *Geography* was very varied and if there was any prominence it was probably in environmental matters and curriculum discussion relating to the Schools Council and DES proposals. Complemented by *Teaching Geography* the Association was now well equipped to meet the needs of the professional geography teacher.

Senior Association losses during this period were again small and confined to Professor Harry Thorpe (President 1974) who died unexpectedly on 14th February 1977 aged 64, and E. C. Marchant, HMI (President 1967) who died 13th September 1979 aged 77.

This was a period of recession in the universities as pressured retirements were not being replaced as a result of a financial squeeze by the Government. Geography, however, was one of the few subjects which managed a certain momentum, albeit reduced, as is witnessed by the appointment of a further fifteen individuals to chairs during the period 1977 to 1980 (see Appendix L for details).

By this time considerable change had taken place in the pattern of school broadcasts. The long running Travel Talks had been signed off in 1965 and replaced by 'Exploration earth', an ambitious project backed up with pamphlets in colour. This was followed in the 1970s by 'Radiovision' with the broadcasts being supported by 35 mm filmstrips. Perhaps taking a hint from *Geography* the BBC also introduced in 1974 a series on 'Our Changing World'. The Geographical Association and academic geographers were brought into all these projects

and this period probably represents a high water mark of radio in the post-war era.

Sadly a relative decline was soon to set in as a result of a combination of factors. Improved recording/replaying facilities in the schools with tapes encouraged a shift of transmission times from day to night time so that 'actuality' programmes disappeared from the school timetable. At the same time TV, offering both sound and pictorial (later colour) resources, was spreading rapidly along with recording/replaying videos. A big reduction in radio time inevitably followed and in 1990 educational broadcasting found itself a minor part of Radio 5.

Only a limited number of TV programmes have been specifically made for school use but astute teachers have realised that a great deal of real geography can be obtained from 'the box' by selective video recording and subsequent editing. The slide projector and film strip has a serious competitor now that the moving picture can be held at a particular point without undue complications and without full black-out of the room.

Although the 1980s were to reveal the importance of the Geographical Association in arguing for a prominent place for geography in the school curriculum the debate took place during anxious times: nationally the early years of the decade were dominated by a recession and continuing inflation, but more serious from the Association's point of view was the fall in school numbers produced by the post-World War 2 fluctuating birthrate. The fluctuating rolls, with an overall national fall of about one third, led to a reduction in the number of teachers and this had a serious impact upon membership numbers. Previous inflationary fluctuations now translated into a steady decline: whereas in the 1970s annual membership, although varying slightly, had always topped 8000, the 1980s saw the figure slip to 7203 in 1980, down to 6635 in 1981, and a low of 6165 in 1982: thereafter there was a slow recovery to 6258 in 1983, 6463 in 1984, 6555 in 1985 and 6520 in 1986. Managing the Association's finances against this background was difficult especially in view of the deficits which had been incurred in 1978 and 1979, but thanks to the vigorous publication programme which had been initiated by the new

Publications and Communications Standing Committee the Association struggled back into the black in 1980 on a turnover of £111,700 and then steadily increased this figure year by year to reach £234,700 by 1986.

There were also growing problems on the political front. The so called 'Great Debate' on education began with a speech by the then Prime Minister, James Callaghan, at Ruskin College Oxford in October 1976. This was followed in 1977 by the DES paper *Educating our Children*, an HMI report *The Curriculum 11–16* (The 'Little Red Book'), and finally by the important Green Paper *Education in Schools: A Consultative Document* which specified eight aims for education. Later that year there appeared DES Circular 14/77 *LEA Arrangements for the School Curriculum* and in 1980 the definitive DES document *A Framework for the School Curriculum*.

The Geographical Association was quick to respond to all these preliminary views as geography was not listed as a core subject, nor given proper recognition in the detailed assessments. The Association quickly prepared and distributed 50,000 copies of a statement *Geography in the School Curriculum 5–16* stating the case for including geography. Although this may possibly have led to some improvement in the thinking at the Department of Education and Science the fact remained that geography was *not* included in the School Curriculum framework issued by DES in 1980 as a mandatory core subject alongside English, Mathematics, Science and Modern Languages despite the fact that it was already the third most popular subject with most O level examining boards.

The situation in geography was also somewhat complicated by many educational theorists largely led by sociologists who wished to replace established school subjects with 'integrated' or cross curricular courses. The 'new ideas' were typically based upon sociology and were called combined studies or integrated humanities. Geography by its very nature was seen to be capable of integration with several subjects, but in particular a combined geography and history was favoured as a 'Humanities' possibility. The danger of geography disappearing altogether from the school syllabus was therefore very real. The

Association was instrumental in mounting the movement to preserve and improve upon the situation which already existed. One of its first important actions taken was the production and circulation in 1980 of the already mentioned leaflets dealing with *Geography in the School Curriculum*: this leaflet stated clearly the contribution which geography was able to make in the overall pattern of education.

Subsequently the Geographical Association submitted written evidence to COSTA (Council of Subject Teaching Associations) which itself had been invited to send a delegation in June 1981 to give evidence to the Parliamentary Select Committee on Education. The next step in the debate was an enlargement in the years from 5–16 to 16–19. Needless to say the intense official and public interest in the overall school curriculum is reflected in the activities and discussions of nearly all the Association committees, working groups and working parties over this period. Both *Geography* and *Teaching Geography* contain numerous articles and references to the progress of the debate during 1977–85 and a valuable summary of the action taken by the Association so far is given by Michael Williams in a paper published in the April 1985 issue of *Geography*.

For the Geographical Association the culmination of the debate came on 19th June 1985 when the Rt. Hon. Sir Keith Joseph Bt. MP., Secretary of State for Education and Science, accepted an invitation to address the Association on the place and nature of geography in the school curriculum. This was the first time that a Minister of the Crown had ever ventured to speak to the Association about geography. The meeting, held in King's College London, attracted a capacity invited audience. In a memorable and stimulating address Sir Keith raised seven wide-ranging questions which were duly answered by the Association. The full details will be found in *Geography* for October 1985 and *Teaching Geography* for January 1986. Sir Keith's address on June 19th was followed by a paper delivered by Trevor Bennetts, HMI, on *Geography from 5–16 A view from the Inspectorate*. This paper outlined the aims and objectives for geography which HMI considered to be appropriate for pupils during different phases of their school education up to the age of 16. The paper was also printed in *Geography* for October 1985.

By now the groundwork for the National Curriculum debate was being effectively laid. Accompanying the debate we find considerable discussion in the profession on the public image of geography: whilst much improved in many respects it was clear from two other projects initiated by the Geographical Association that there was still room for much more work. The first exercise was the World Wise Quiz competition which began in 1984 as a result of a proposal made by Michael Morrish and Rex Walford in May 1983. Organised initially each year by the Branches on a regional basis the finals have followed at the Annual Conference. Several hundred schools have participated on each occasion and most of the book publishers as well as the Field Studies Council and the Ordnance Survey have supplied appropriate prizes at various stages of the competition.

The second exercise promoted by the Geographical Association was the Map Watch of November 1984. Correspondence between the Geographical Association and the BBC regarding the quality of the Weather Maps on TV led the Educational Standing Committee to propose that November 1984 be declared a Map Month in which members would be invited to monitor the quality of maps in the media. Professor Balchin was invited to organise the collection and analysis of the data. Several other societies with cartographic interests joined in the exercise and many teachers enthusiastically adopted the Watch as a project. As a result several thousand maps and reports were submitted by several hundred participants. A full report of the analysis was published in the October 1985 issue of *Geography*, after a one day seminar in Cartography for the Media had been held at the London School of Economics on Wednesday, 10th July 1985.

The rapid turnover of voluntary officers resulting from the 1977 revision of the constitution continued into the 1980s and life in the Association became characterised more by team work rather than individual action. After Professor Mead (1981–82) the Presidents were Professor Lawton (1982–83), Rex Walford (1983–84), Pat Cleverley (1984–85), Patrick Bailey (1985–86) and Professor Brunsden

(1986–87). The Honorary Secretaries were B. E. Coats (1979–84), M. T. Williams (1981–85), Elspeth M. Fyfe (1984–87) and J. A. Binns (1985–89). Assisting these as Honorary Treasurer were F. E. Hamilton (1981–84) and B. E. Coates (1984–87); as Library and Information Officer T. W. Randle (1977–83), Elspeth M. Fyfe (1983–85) and M. J. Shevill (1985–91); as Trustees J. Old (1977–86), F. E. Hamilton (1981–84), Professor Patmore (1979–88), B. E. Coates (1984–87), and Professor Stanley Gregory (1986–90); as Publications Officer D. J. Boardman (1980–86) and Patrick Weigand (1986–91).

The Editors of *Geography* were R. J. Small (1980–83), Russell King (1983–89) and of *Teaching Geography* Patrick Bailey (1975–85) and Eleanor Rawling (1985–88). The Honorary Conference Organisers were B. S. Morgan (1979–82), K. Hilton (1983–84), D. R. Green (1985–86) and N. Yates (1987–89). Changes also took place with the paid administrative staff as Mr. R. Fell found it desirable to retire three years early on 30th September 1981. He was replaced by Mrs. Susan J. Pitt who in turn resigned in 1984. The next occupant of the post was Miss Anne Courtenay (later Mrs. Anne Argent) who resigned in 1986, to be followed, as Senior Administrator, by Miss Margaret Barlow from 1986 to 1990 and Miss Frances Soar from 1990.

Council over this period was heavily committed to responding to the various DES documents, more especially a *Framework for the School Curriculum* which prompted the production and circulation of the already mentioned leaflets on 'Geography in the School Curriculum 5–16'. The problem of the Fleure Library was resolved for the time being by the University of Sheffield Library taking over the responsibility for its care and maintenance in 1982: this solution released space at Headquarters which was subsequently taken over by the Sheffield Academic Press.

A Board was established by Council for the validation of in-service education for teachers (INSET) in 1983; additionally Regional Conferences were organised in connection with the curriculum debate; November 1984 was selected as Map Month for the national monitoring exercise; approval was given for the institution of Associate Membership; a response was prepared to Sir Keith Joseph's address to the Association; a GA International Relations Fund and a GA New Initiative Fund were established and assistance provided in the construction of the Council of British Geography which was formed as a result of the demise of the Royal Society's National Committee for Geography.

Additional Honorary Members were created in 1981 (Professor S. H. Beaver, Mr. L. J. Jay and Mr. J. A. Morris), in 1982 (Professor R. W. Steel), in 1983 (Mr. R. S. Barker, Professor Norman Pye, Professor M. Wise), in 1985 (Miss Rosemary Robson) and in 1986 (Lord Nathan).

The early 1980s saw intense activity in both Standing Committees of Council. Education Standing Committee by now had not only its four historic section committees but also five working groups which by 1984 had spawned no less than seven working parties. It is impossible in a condensed account to detail all the work undertaken but in addition to the ongoing standard diet of conferences, curriculum discussion, examinations, relations with the TV and radio world, field work and career opportunities, the committees and groups also considered a diverse list of topics such as multicultural education, the revision and supply of school textbooks, Schools Council memoranda (until its closure in 1984), ancillary assistance in schools, the impact of the Health and Safety Act, the Warnock Report on children with special needs, the effect of fluctuating rolls, the public image of geography, map work, conservation strategies, the 1986 computer-based Domesday project and managing a Geography Department.

The Publications and Communications Standing Committee was equally active with a stream of publications. The already noted Secondary School Handbook by Professor Norman Graves appeared in 1980 to be followed by a Primary and Middle Schools Handbook edited by David Mills in 1981. Other publications included *Geography into the 1980s* by Eleanor Rawling, *Computer Assisted Learning in Geography* (jointly with the Council for Educational Technology), *Patterns on the Map* (dealing with the Second Land Utilisation Survey), several Landform Guides, *The Good, the Bad and the Ugly*, *Geography with Slow Learners*, *Teaching Geography to the Less Able*, *Geographical Futures*, *Designing and Teaching Integrated Courses*, *Evaluation and Assessment in Geography*, *Geographical Education for a Multi-Cultural Society*, *Geography, Schools and Industry*, *Perspectives on a Changing Geography*, *Geography Beyond A level*, *A Handbook for Geography Teachers*, *The Role and Value of New Technology in Geography*, *Profiling in Geography*, *Geography and Careers*. This is a remarkable record of purely voluntary self-help for the profession as neither authors nor editors received any financial recompense for their labours. With this major effort it is not surprising that by the middle of the decade annual sales of publications were ten times that of the beginning and the programme was a critical factor in pulling the Association out of the red and overcoming the problems produced by falling school rolls and membership of the early 1980s.

The work of the Council and its Standing Committees produced an immense load of work for the Headquarters staff and the rapid turnover of Senior Administrators (4 changes in 5 years) did not help matters; much credit is due to a core of juniors who kept the Association running smoothly whilst endeavouring in addition to master the complexities of computerisation of all membership records.

The Annual Easter Conference during the 1980s increasingly became the high spot of the Geographical Association's national activities as attempts to mount a plenary summer gathering failed. Smaller specialist summer seminars and one-day sectional conferences were, however, popular and much useful work was accomplished thereby. The Annual Conferences, however, were not without problems as the London School of Economics which had traditionally provided free accommodation was forced to charge for the hire of the premises, steadily increasing the fees in line with inflation. As a result the Annual Conference ran into a deficit for the first time ever in 1983. The situation was rapidly rectified by adjusting the cost of advertisement space in the exhibition, experimenting with an alternative venue in King's College London, and also arranging one of the days at the Royal Geographical Society. The Geographical Association Annual Conference thus maintained its open door policy to all geographers at no cost. The Association must be unique nowadays in not charging any conference fee for attendance.

Attempts to organise overseas summer excursions were also casualties due partly to the recession and partly to the increase in package holidays. Both a West Indies (1983) and a West Coast of America (1985) proposal were floated but failed to materialise and only a Thailand (1982) excursion with 16 participants took place. This was a far cry from the heady days of 1970 when over 300 applications were received for the Association's memorable Trans-Canada Study Tour.

Branch life suffered with the recession and in some cases lapsed as active officers retired: however some 64 groups, including new or resurrected branches at Walsall, Northants, Durham, Cambridge, Hereford, Hertfordshire, Waltham Forest, Watford, Aylesbury, Ambleside and Shrewsbury maintained a varied programme of activities. One unfortunate trend was a decline in the popularity of field excursions—which seems to correlate with the spread of car ownership. The changing social conditions became a challenge to Branch officers to devise unusual group excursions such as day flights to Paris, hovercraft trips to Boulogne and river cruises on the Thames and Cherwell.

Both *Geography* and *Teaching Geography* during this period reflect the growing concern with the curriculum and there are few issues without reference to the debate. The main sources have already been mentioned above. *Geography* under the editorship of R. J. Small (1980–85) maintained the balance between longer home and overseas articles, with shorter 'This Changing World' contributions, together with the standard reviews and news of the Association. With the editorship of Russell King from 1984 onwards however the familiar red cover with its Mollweide world projection was replaced by a space age cover and a new feature under the title of Geo Notes appeared. These covered items of interest in the field of education, remote sensing, OS developments, computing, and media maps which did not easily fit into the format of 'This Changing World'.

*Teaching Geography* continued under the editorship of Patrick Bailey (1975–85) with a

steady flow of articles for the classroom teacher: map work and field work inevitably figured prominently whilst the journal also reflected a growing interest in the profession with studies of natural hazards. A new venture began in 1985 with the appearance of a *GA Newsletter* under the initial editorship of Michael Morrish.

Losses amongst the senior retired members were again few and confined to the deaths of Mr. A. D. Nicholls (Honorary Member and Past President 1972) in 1981, Professor K. C. Edwards (President 1963) in 1982, Professor Stanley Beaver (Honorary Member and Past President 1966) in 1984, Leslie J. Jay (Honorary Member and Honorary Librarian 1955–73) in 1986, and Professor J. A. Steers (President 1959) in 1987. All were stalwart members of the Association and Professors Edwards, Beaver and Steers contributed greatly to our knowledge of the geography of the British Isles. Professor Steers' interest in coastal physiography led to a unique record as he had walked the entire length of the coasts of England and Wales and also much of that of Scotland. His expert knowledge was a major factor in the creation of the Nature Conservancy and the National Parks in the 1940s.

Falling school rolls had begun to affect the universities by this time and little change took place in the professoriate owing to the severe Government financial cutbacks.

### The National Curriculum Debate

The concluding years of the Geographical Association's century of existence have been dominated by the emerging concept of the need for a National Curriculum to provide an equivalent foundational education for all. Although over the years the examination boards had evolved a kind of national curriculum for some subjects there inevitably remained many anomalies which made things difficult when pupils moved school. There was also increasing public concern with the so-called progressive methods of teaching whilst geography and history had their own 'Humanities' problem.

In 1987 the HMI geographers produced an authoritative statement on the value of geography under the title of *Geography from 5–16 Curriculum Matters 7.*

This was an uncontroversial statement which countered the ideas of the sociologically orientated 'progressives' as it placed the emphasis on facts rather than debate and discussion of controversial issues. During the same year Sir Keith Joseph was replaced by Kenneth Baker as Secretary of State for Education and Science and the Association immediately resolved to continue the debate from the point reached with Sir Keith. The Association's more substantial response to Sir Keith's seven questions was speedily completed and sent to Mr. Baker. It was then published under the title *A Case for Geography.* This statement proved to be a significant step in confirming the place of geography in the school curriculum. Mr. Baker invited the Association to meet him on June 30th 1987 when he indicated that geography would be regarded as a foundation subject in its own right in the curriculum and that a Geography Working Group would be brought into existence in 1988 after the groups for the 'core' subjects of English, Maths and Science had reported.

A Geographical Association Shadow Working Group was immediately set up in anticipation of official action, and a discussion paper, *Geography in the National Curriculum,* was prepared to give the Association's view of the kind of geography it would like to see in a national curriculum: this was subsequently published in March 1989. Meanwhile the expected official action had taken place with the passing of the Education Reform Act of 1988: this provided for the establishment of a National Curriculum comprising core and foundation subjects to be taught to all pupils of compulsory school age in maintained schools.

At about this time a Council of British Geography emerged as a result of the demise of the Royal Society's National Committee for Geography (abolished along with all other national committees for financial and other reasons). The new Council is made up of representatives from all the geographical bodies including the Geographical Association. So far it has played a somewhat limited role in the National Curriculum debate but it has the potential to exert a considerable influence in the future.

To return to the Education Act of 1988: proposals as to the core subjects and supporting foundation subjects gradually emerged from the DES. Although there were welcome signs that the importance of geography was at last being realised the profession was disappointed that it did not appear as a core subject along with English, Mathematics and Science despite the clear evidence of its importance in the GCSE examinations (Table Fig. 7). There was no doubt however that geography was now a foundation subject in the school curriculum: a state of affairs for which the Association could claim much credit.

The first National Curriculum Working Group to report was Science: its Interim report immediately revealed an overlap with traditional geography teaching. Large elements of earth science and people-and-environment studies were included. The Geographical Association immediately took up this specific problem of overlap with the Secretary of State and a DES committee was quickly established to oversee overlaps in all subjects.

The next stage in the saga was the setting up of a National Curriculum Working Group for Geography. Early in 1989 the Association made suggestions to the DES as to the composition of the Group and when details were announced it was found to contain (perhaps not surprisingly) no less than nine well known and active Association members. The group was appointed in May 1989 and the Association immediately despatched briefing papers. An Interim report for discussion appeared in November 1989.

The Association then devised a strategy to produce a considered response from its membership by arranging for thirteen regional conferences in January 1990: the results of these conferences were combined with the Council's reactions and forwarded to the Secretary of State. The final report *Geography for Ages 5–16: Proposals of the Secretary of State for Education and Science and the Secretary of State for Wales* was published in June 1990. This was not the end of the story however as further discussion was still possible. The Association accordingly organised a further sixteen countrywide regional conferences: Altogether over 3,000 members participated to

**Table 1**
**Joint Council for the GCSE: 1988 examinations**

| Subject | Number of candidates | The following figures show the provisional results statistics in national criteria subjects: Percentage of candidates obtaining grade | | | | | | |
|---|---|---|---|---|---|---|---|---|
| | | A | B | C | D | E | F | G |
| Art and design | 221,438 | 8.8 | 12.5 | 18.9 | 18.8 | 17.5 | 14.5 | 7.8 |
| Business studies | 22,741 | 6.2 | 9.2 | 22.8 | 22.9 | 16.4 | 10.8 | 5.1 |
| Latin | 15,609 | 41.8 | 26.1 | 17.3 | 7.4 | 4.0 | 1.9 | 0.9 |
| Greek | 1,276 | 62.6 | 23.2 | 9.8 | 2.5 | 1.0 | 0.6 | 0.16 |
| Classical civilisation | 3,066 | 17.0 | 21.0 | 22.4 | 9.8 | 7.4 | 6.9 | 6.7 |
| Computer studies | 97,028 | 5.5 | 11.0 | 17.6 | 16.9 | 16.3 | 15.6 | 11.9 |
| Craft, design, tec. | 157,021 | 6.8 | 11.4 | 16.9 | 19.0 | 18.8 | 15.4 | 7.6 |
| Economics | 32,449 | 9.7 | 15.1 | 22.2 | 20.7 | 16.4 | 9.3 | 4.3 |
| English | 654,742 | 6.6 | 14.2 | 22.8 | 23.2 | 17.9 | 10.3 | 4.0 |
| English literature | 394,260 | 9.4 | 17.3 | 24.1 | 22.3 | 15.5 | 8.0 | 2.6 |
| Home economics | 187,543 | 4.1 | 10.0 | 19.1 | 21.2 | 20.0 | 15.1 | 7.3 |
| Geography | 295,163 | 7.8 | 14.0 | 19.5 | 19.1 | 16.4 | 12.5 | 7.3 |
| History | 242,760 | 9.8 | 15.0 | 19.3 | 17.4 | 15.5 | 11.9 | 7.6 |
| French | 238,132 | 19.8 | 14.6 | 15.6 | 17.9 | 14.8 | 11.6 | 4.7 |
| German | 68,675 | 20.7 | 15.3 | 16.7 | 17.5 | 13.6 | 10.9 | 4.4 |
| Spanish | 17,448 | 22.9 | 17.4 | 16.8 | 15.9 | 12.0 | 9.9 | 4.5 |
| Mathematics | 661,655 | 6.2 | 8.9 | 21.7 | 16.4 | 16.3 | 15.4 | 7.1 |
| Music | 27,577 | 13.3 | 21.2 | 23.2 | 15.7 | 11.9 | 8.2 | 3.3 |
| Religious studies | 104,009 | 6.8 | 12.9 | 18.8 | 17.7 | 16.8 | 12.6 | 7.7 |
| Biology | 293,949 | 7.3 | 12.4 | 23.1 | 19.2 | 14.4 | 12.2 | 8.5 |
| Chemistry | 214,818 | 10.6 | 14.7 | 23.0 | 18.7 | 14.0 | 10.8 | 5.2 |
| Physics | 245,218 | 9.1 | 14.5 | 21.1 | 19.7 | 16.9 | 12.0 | 5.8 |
| Science | 113,091 | 4.6 | 8.1 | 15.8 | 21.5 | 21.6 | 17.5 | 9.0 |
| Social science | 10,999 | 4.1 | 8.1 | 15.0 | 13.9 | 17.7 | 19.3 | 12.2 |

The above figures refer to Mode 1 only. Mode 1 describes exam papers set by the examining boards without teacher involvement. It covers 86 per cent of all papers.
SOURCE: *The Guardian*, 25th August 1988.

Figure 7. Geography in the GCSE 1988 examinations.

consider the report and to make recommendations. The next step was submission to the National Curriculum Council and then return to the Secretary of State for further comment before draft orders were laid before Parliament in the Spring of 1991.

It was thought by many in the profession that the Interim report was a masterly attempt at solving a difficult problem: it even received a rapturous welcome in a *Times* Editorial (Fig. 8).

Others however were somewhat dubious in view of the very large factual content and perceived conservative approach. Either way the report concentrated upon basics and brought back a proper element of physical geography into the subject, whilst at the same time endeavouring to preserve as much freedom as possible for teachers to introduce their own special interests and expertise into their courses. Subsequent official action largely consisted of chipping away desirable sections of the report as it was realised that there was not enough time in the school week for the core subjects and all the foundation subjects. The problem was also compounded because of the overlaps which emerged with the interim reports in Mathematics and Science: the Association found it necessary to respond to these overlaps, especially those in physical geography.

The overload aspect was most acute at Key Stage 4 (14–16 age group) and a new Secretary of State for Education, the Rt. Hon. Kenneth Clarke (appointed in the Cabinet reshuffle which followed Mrs. Thatcher's resignation) announced on January 4th 1991 that only the core subjects would remain obligatory and that schools would be allowed flexibility in selection from the foundation subjects, with three possibilities open to geography and history at Key Stage 4: either single subject courses or a combined joint course could be taken. This was seen by both subjects as a step backwards and

# NOT JUST ABOUT MAPS

Geography, says the dictionary, is "the study of the natural features of the earth's surface . . . and man's response to them." The dictionary speaks true. Geography embraces every *fact* on earth: every aspect of the composition, occupation and history of the planet. It is the monitor of our abuse of our environment and our guide to its preservation. As such, geography knows no intellectual boundaries. It deserves to sit at the centre of any liberal education.

School geography has none the less recently had to fight its way back from being a mere option to being one of the foundation subjects in the Government's National Curriculum and is still far from being one of the "core" subjects. While the grandees of English, science and mathematics sit luxuriating above the salt, geography was originally left to fight it out with history as an option for teaching time, below even such peripherals as French and gym. Now, geography will at least become compulsory. But its lowly status is an educational outrage, a comment on the domination of teaching in Britain by the universities and their medieval academic priorities.

Yesterday, the Government's working party on geography began what could be a long rearguard action. The new, flexible list of guidelines for teachers of children at ages 7, 11, 14 and 16 shows that geography's small band of guerrillas has bravely captured much of the "green" territory now so fashionable among school children − and thus being coveted by the conservatives for "pure" science. In future, seven-year-olds will be expected to know about the weather, about their neighbourhood, about travel and land use. By 14, they should know about the configuration of the landscape, its impact on population, industry and transport, about the food chain and species survival. By 16, pupils will have strayed more confidently into soil science, economic history and the regulation of the environment. They should be able to recognize the world about them and understand the natural and human forces which shape and change it.

Yet these are no more than spoils from the outer bailey of the "core" curriculum. Geography should be encouraged to seize the central fortress, ejecting both pure science and that grossly over-promoted intellectual exercise called mathematics. Geography should stand alone on the scientific pedestal, joined only with its one educational equal, the study of the human spirit in English language and literature. Geography is queen of the sciences, parent to chemistry, geology, physics and biology, parent also to history and economics. Without a clear grounding in the known characteristics of the earth, the physical sciences are mere game-playing, the social sciences mere ideology.

The education secretary, Mr John MacGregor, said yesterday that geography was vital for pupils to gain "an informed appreciation of the world in which they are growing up and in which they will live and work as adults." Nobody would quarrel with that. But why does a government so commendably interventionist in matters curricular not put its words into action? Why does it traipse along behind the academic conservatives? If Mrs Thatcher's "full repairing lease" on the earth is to be honoured, British children must be taught how to do-it-themselves. Geography should be declared a core.

Figure 8. *The Times* Editorial of 7 June 1990.

remains a partially unresolved problem at the time of writing.

On January 14 1991 the Draft Order for Geography was published. This again produced a reaction as it proposed a major reduction of approximately one third of what schools would be required to cover compared with the Geography Working Group's recommendations. Even more emphasis was placed upon basics and simplification: but there did seem to be a greater freedom for teachers to develop their own courses. The Statutory Order provides quite specific details of a simplified content to be covered in five Attainment Targets and outlines a set sequence of places to be studied in the Programmes of Study, and refers to ten levels of achievement in relation to specific knowledge and skills to be covered.

The major gain in the whole National Curriculum exercise has been the formal recognition of geography as a foundation subject plus the decision to re-establish the subject firmly in the 23,000 Primary schools of England and Wales after its shadowy existence or disappearance in these schools in the post World War 2 period. Geography must now be part of the curriculum at Key Stages 1, 2 and 3 (ages 5–7, 7–11, 11–14) for all pupils in maintained schools. The confirmation of the importance of physical geography is also to be welcomed: only the problem of an overcrowded curriculum at Key Stage 4 remains to be resolved.

Comparing these curriculum discussions with those at the beginning of the century the most conspicuous omission in the new look is perhaps 'General World Geography': but doubtless the ingenuity of many teachers will overcome this problem within the permitted flexibility.

The National Curriculum in Geography was formally 'launched' at a Conference in London organised by the Royal Geographical Society on 24th September 1991. The keynote address was given by the Rt. Hon. Kenneth Clarke, QC MP, then Secretary of State for Education and Science, who emphasised the importance of geography at all levels of the educational process. Basic facts, locational geography, the study of places, primary geography, physical geography and geographic skills all received particular attention, and mention was also made of the unresolved Key Stage 4 problem. The Secretary of State was followed by Sir Leslie Fielding, KCMG, Vice Chancellor of the University of Sussex and Chairman of the National Curriculum Geography Working Group who gave a background talk on the work of the Group. Other contributions came from Association members, all of whom were seeking ways of making the National Curriculum work. Kay Edwards explained how she and her colleagues had planned, designed and implemented Key Stage 3. Richard Daugherty tackled the difficult question of assessment. Wendy Morgan considered the challenge presented to non-specialist teachers in primary schools whilst Eleanor Rawling summarised the ways in which the Association was supporting members in implementing the National Curriculum.

On the morning after the Conference *The Times* ran a critical Editorial concerning geography and this provoked further correspondence including a riposte from the Secretary of State. A report of the Conference will be found in the *Geographical Journal*, Vol. 158, Pt. 1, for March 1992 whilst Kenneth Clarke's address is also available in *Teaching Geography* for January 1992.

One consequence of the National Curriculum discussion on geography has been the emergence of a separate Association of Geography Teachers in Wales to cope with the particular problems of the Principality. A separate report had to be prepared for Wales because of the bi-lingual problem. With a small number of schools all Welsh, others bilingual, and a majority all English, it follows that timetable problems abound: whilst the addition of Welsh as a subject for study in many schools means further pressure on already congested timetables.

With Parliamentary approval given to the Statutory Order for Geography the main phase of the Association's National Curriculum work (apart from Key Stage 4 for the 14–16 age group) has, for better or worse, been completed. The main thrust of the Association's activities for the mid 1990s will be to provide positive support for geography teachers in implementing the curriculum. This process has already begun with the issue of leaflets and notes to assist at Key Stages 1, 2 and 3 whilst a programme of more substantial publications is in hand and should be available by the time this account appears in print.

The need to prepare for and respond to the National Curriculum operation (over 19 documents were produced) combined with the normal work of the Association placed an enormous load of work and responsibility on the senior officers of the Association during this period. After Professor Denys Brunsden (1986–87) the Presidency was filled by Dr. Graham Humphrys (1987–88), Mr. Michael Storm (1988–89), Mr. Richard A. Daughterty (1989–90), Mr. Bryan Coates (1990–91), and Mrs. Eleanor M. Rawling (1991–92) with the centenary year being covered by Mr. Simon Catling (1992–93) and Professor Andrew Goudie (1993–94). The Honorary Secretaries were J. A. Binns (1985–89), P. Fox (1987–93), and D. Burtenshaw (1989– ). Assisting as Honorary Treasurer were P. J. M. Bailey (1987–88) and N. Simmonds (1988– ), and as Trustees S. Gregory (1986–90), P. J. M. Bailey (1987– ), N. Simmonds (1988- ) and R. A. Walford (1990– ). The Publications Officers were P. Weigand (1986–91) and Dr. J. Hindson (1991– ) with the Library and Information Officers M. J. Shevill (1985–91) and K. Luker (1991– ).

Council during this period additionally dealt with a Report to the Royal Society on the teaching of physical geography, made a submission to the Higginson Committee on the A level examination, supported the setting up of the Council of British Geography in 1988, approved new initiatives on Primary membership, considered suitable activities for the centenary celebrations in 1993, discussed further the possible shortage of geography teachers in the 1990s, arranged for a visit to Poland of 15 members in April 1989, processed the launch of a new magazine for primary circulation in 1989 (the advent of the National Curriculum either triggered or coincided with the perceived need to clarify and consolidate the geographical education provided for young children), and introduced NERIS (the National Educational Resources Information Service) with the aid of a £31,000 grant from the Department of Trade and Industry.

In 1990 Council endorsed the Geographical responses to the Interim and Final Reports of the official National Curriculum Geography Working Group and also approved responses to all other National Curriculum subject proposals. It also campaigned vigorously on the crucial issue of geography in Key Stage 4. Council also co-operated with the new Association of Geography Teachers in Wales and the Association Branch in Belfast in respect of the unique curriculum problems in Wales and Northern Ireland, Furthermore it embarked upon a revision of the constitution and created a Strategic Planning Team to look to the future and make proposals on the direction that the Association should take.

The three Standing Committees of Council were also much involved in the National Curriculum saga on top of their normal activities. The Education Standing Committee supported by its four section committees, seven working groups and six working parties additionally considered INSET provisions, moderation in GCSE examinations, variations of the A level syllabus, questions of safety and funding in field work, the level of examination fees, the supply of geography teachers in the 1990s, established a new working party on GIS (Geographical Information Systems) and organised a number of specialist conferences.

The Publications and Communications Standing Committee was responsible for another wide ranging set of publications including *Computers in action in the Geography Classroom, Starting to Teach Geography, A Case for Geography,* Fieldwork Location Guides (several), Classic Landform Guides (several), Worldwise Quiz Books (several), *Managing the Geography Department, Methods of Presenting Fieldwork Data, Geography in the National Curriculum*, and *Geography Through Topics in Primary and Middle Schools.* Perhaps the most notable event however was the launch of a new journal, *Primary Geographer,* under the editorship of Wendy Morgan in the spring of 1989: this became an instant best seller clearly filling a gap in the journal market. Publication followed the introduction of a Primary Day in the Annual Conference of 1988. The P and C Committee also introduced in 1988 an annual Award Scheme to recognise products which have either made, or are likely to make, a significant contribution to geography at all educational levels.

The Worldwise Quiz competition continued its rise to fame and by 1990 over 500 schools were participating in over 50 local rounds in England and Wales and Northern Ireland. In addition the Scottish Association of Geography Teachers organised a series of local rounds in Scotland and the winners of the Scottish final joined the winners of the eight regional finalists to compete in the grand final during the Annual Conference. By this time this type of competition had also spread overseas to geographers in Canada and the USA.

Both *Geography* and *Teaching Geography* throughout this period reflect the discussions organised over the content of the National Curriculum and of course report the progress of events: both however continued with their normal features. The editorship of *Geography* changed in 1990 from Professor Russell King to Dr. D. J. Spooner, innovations after 1988 include thematic presentations of This Changing World (Famine in Africa, Forty Years of the People's Republic of China, the Changing Face of Eastern Europe and Transport in the United Kingdom were notable examples) and there is a much improved space-age cover after January 1990. There was a slight improvement in the number of contributions in the physical field but the editorial plea for more material in this sector continued. *Teaching Geography* also changed editors in 1988 from Mrs. E. M. Rawling to Mrs. J. Kelly and in 1990 from Mrs. J. Kelly to Dr. D. Boardman. Innovations in *Teaching Geography* included special issues on Primary teaching and the National Curriculum, and the introduction of much appreciated photo resource material in full colour. As already noted *Primary Geographer* under the editorship of Wendy Morgan made a good start in 1989 whilst the Newsletter continued under the editorship of Michael Morrish until June 1989 when it changed to Bob Leake.

Whilst Finance and General Purposes was able to report continued satisfactory progress with a turnover rising to £316,215, record sales of £84,276 and assets of £363,378 in 1989, membership figures showed a slight decline from 7191 in 1987 to 7044 in 1988 and 6873 in 1989. The launch of *Primary Geographer* in 1989 however brought well over 1000 new members and the membership figure for 1990 amounted to 8499. The Honorary Treasurer changed in 1988 from Patrick Bailey to N. M. Simmonds and among the Trustees Rex Walford replaced Professor Stanley Gregory in 1990.

The high spot of the Association's national activities continued to be the Annual Conference with its unique National Publishers' Exhibition, Worldwise Quiz competition and varied programme of lectures, seminars and workshops. Until 1990 the Easter Conferences were organised on the basis of a day at the Royal Geographical Society and two days at the London School of Economics and attendances of well over a thousand were common. A break with tradition occurred in 1991 however as it was decided to hold the Annual Conference and the AGM at UMIST in Manchester. The Honorary Conference Organisers were N. Yates (1987–89) and R. Chapman (1990– ).

Branch life continued along traditional lines with temporary closures being balanced by new or resuscitated branches, the total fluctuating between 65 and 70 yearly. The Worldwise Quiz competition provided a welcome new activity for many branches which undertook the initial eliminating rounds: branches were also active in the drive for Primary recruits.

Sadly the Association lost two stalwart members during this period. In 1988 Rex A. Beddis died suddenly at the age of 56. He had served on the Secondary Schools Section committee and was a member of Council throughout the 1970s, and was Publications Officer from 1976 to 1980. Another grievous blow was to follow on 5th March 1989 when Professor Alice Garnett died at the age of 85. The Association lost not only a staunch friend but also one of the last links with its founders, most of whom were personally known to Professor Garnett. As will be recollected from these pages Professor Garnett was Honorary Secretary from 1947–67, President in 1968, Chairman of Council from 1970–73 and an Honorary Member from 1980: with the sole exception of Professor Fleure the Association probably owes more to Professor Garnett than any other member for its success in the present century: sadly she has just failed to make the centenary which she was anticipating with so much pleasure.

The successful Annual Conference held in Manchester in April 1991—the first outside London in the post WW2 era—encouraged Council to hold its successor in April 1992 in Southampton, appropriately under the title of 'Changing Curriculum—Changing Geography'. The Conference again attracted a record attendance of both exhibitors and participants and will probably be long remembered for an outstanding address by Simon Jenkins, then Editor of *The Times*, who provided *en passant* an explanation of the support which the paper had been giving in recent years to the cause of geography. The Presidential Address from Eleanor Rawling was also memorable as a critical assessment of the National Curriculum for Geography revealed a number of deficiencies which will need eventual correction. Teachers are already finding the implementation of parts of the National Curriculum difficult and non-specialists in particular will need the assistance of the Geographical Association. The assessment of pupil achievement raises many new problems and it will be some time before the system is working smoothly. The need to maintain links with academic geography was also stressed together with a flexible interpretation of the curriculum in order to absorb new ideas in the future and so avoid fossilisation in the schools. There are thus already murmurs of desirable amendments to the Geography Order and it is clear that curriculum discussion will continue well into the '90s.

Council reported for 1991 its critical response to the Geography Draft Order which was taken note of in the final Statutory Order approved by Parliament in March 1991. Attention then turned to supporting the creative implementation of the National Curriculum with a major Geographical Association programme of publications and regional seminars. Key Stage 4 (14–16) and its problems also received much attention. The first Geographical Association validated Diploma was awarded in May 1991 to Edge Hill College of Higher Education. A report from the Strategic Planning team gave rise to considerable discussion as to future Association objectives and a more open method of electing officers by postal ballot was approved. Professor Norman Graves was made an Honorary Member.

The Education Standing Committee devoted many hours to responding to a multitude of documents concerning geography that were issued in connection with the curriculum consultation. A series of 10 Regional Inset Seminars were approved and these provided successful ways of exploring opportunities for the Association to supply effective in-service support for the future. The Working Groups were all asked to produce Action Plans and the mid 1990s should be marked by some lively discussions.

The Publications and Communications Standing Committee also reported a busy year. The flow of publications continued (see Appendix M), the Worldwise Quiz attracted entries from over 600 schools and there were 70 local rounds in England, Wales and Northern Ireland with 11 in Scotland. The value of the sponsorship for this competition was estimated to have reached £23,000. The success of *Primary Geographer* was also a particularly pleasing aspect of the year.

Finance and General Purposes Standing Committee also reported on an optimistic note as it was possible to record a modest surplus at the year end, but income and expenditure remain very finely balanced. The income/expenditure turnover was of the order of £350,000 for the year with assets at some £367,000. Total membership was 9672, an increase of 1184 on the previous year: this was welcome news but the increase was wholly made up by a rise in the primary membership and there was a slight decrease in secondary teachers—a situation which will need careful monitoring in future.

One problem which the Association now faces is the fluctuating nature of its finances as a result of the partial dependence upon income from publications. Publication of books, pamphlets, etc., involves initial capital investment, the return coming in subsequent years. It is not easy to maintain a steady flow of new publications, especially with changing personnel, so the financial statements now tend to show a recurring surplus/deficit/surplus pattern, Fig. 9, Comparative Statistics, reveals this on a steadily rising turnover. A long term solution to the problem might be the creation of a separate publication account, building up an initial capital fund as a first step.

As the centenary approaches probably one of the most important steps taken by the Association in the flurry of activity surrounding the National Curriculum debate will be seen to have been the publication of *Primary Geographer* and the insertion of a Primary Day in the Annual Conference to serve the needs of the teachers in the 23,000 primary schools who now find that Geography *per se* has become

**Comparative Figures**

| Year ending 31 August | 1987 £ | 1988 £ | 1989 £ | 1990 £ | 1991 £ |
|---|---|---|---|---|---|
| **INCOME** | | | | | |
| Subscriptions[1] | 144,957 | 153,675 | 155,339 | 189,267 | 210,022 |
| Sales of Publications[2] | 37,558 | 34,055 | 84,276 | 57,682 | 71,246 |
| Royalties | 250 | 88 | 1,653 | 1,133 | (356) |
| Advertisements & Inserts[3] | 14,396 | 14,106 | 13,887 | 15,289 | 16,119 |
| **EXPENDITURE** | | | | | |
| Journals | | | | | |
|   Printing & Publishing[4] | 32,753 | 34,053 | 37,103 | 53,720 | 55,836 |
|   Editorial Costs | | | 1,659 | 3,050 | 1,533 |
|   Postage & Despatch | 13,741 | 16,622 | 14,521 | 15,533 | 16,351 |
| Special Publications[5] | 12,223 | 12,543 | 28,504 | 28,506 | 15,955 |
| Salaries, NI, Superannuation[6] | 73,223 | 80,395 | 88,461 | 101,657 | 116,551 |
| Audit, Banking & other Professional Charges[7] | 2,039 | 3,479 | 3,461 | 3,566 | 5,768 |
| Officer & Committee Expenses[8] | 7,538 | 9,007 | 9,938 | 12,076 | 14,340 |
| Establishment Costs | 18,065 | 17,646 | 21,066 | 23,795 | 24,487 |
| **STOCK** | | | | | |
| Special Publications | 26,221 | 40,873 | 34,365 | 33,026 | 43,368 |
| **INVESTMENTS (AT COST)** | 203,055 | 201,245 | 244,997 | 265,086 | 263,457 |
| **INVESTMENT INCOME[9]** | 36,528 | 35,134 | 33,223 | 30,482 | 25,578 |
| **PROMOTION OF PRIMARY GEOGRAPHER[10]** | – | – | 13,416 | 2,630 | 1,380 |
| **TURNOVER[11]** | 247,621 | 263,895 | 316,215 | 326,177 | 365,706 |
| **RESERVES** | | | | | |
| HQ Replacement | 100,000 | 125,000 | 148,000 | 148,000 | 153,000 |
| HQ Maintenance | 16,500 | 17,000 | 17,000 | 17,000 | 17,000 |
| General | 118 000 | – | – | – | – |
| Accumulated | 77,783 | 197,114 | 198,378 | 195,829 | 195,866 |
| **SURPLUS/(DEFICIT)[12]** | 36,274 | 26,831 | 26,264 | (2,549) | 5,037 |
| **SURPLUS/(DEFICIT) ON OPERATIONS** | (254) | (8,303) | 6,457 | (30,401) | (10,411) |
| **MEMBERSHIP (numbers)[13]** | 7,191 | 7,044 | 6,873 | 8,488 | 9,672 |

**Notes**

1  The increase mainly reflects a continuing growth of Primary Membership which masks a slight fall in membership in other categories.
2  Publications sales were encouragingly buoyant and provided a welcome increase to the Association's income.
3  Advertisements and inserts remain a little sluggish in times of recession.
4  Printing and publishing costs have been held very tightly in check.
5  The apparent downturn results from expenditure already made in the previous year and not from a reduction in the flow of publications.
6  The Association is linked to the University awards and the increase reflects this.
7  Extra payments made this year to establish the system of computerised accounts and to clarify the Association's VAT position.
8  Extra expenditure was incurred for meetings about such matters as the Strategic Review, Constitution Working Party and National Curriculum.
9  The sharp fall in interest rates has adversely affected investment income.
10  This only covers the easily identifiable costs of leaflet printing.
11  Turnover increased by just over 12%.
12  A small overall surplus is recorded and a considerable reduction on the operational deficit from 1988–90.
13  The increase in membership total reflects the take up of Primary membership.

Figure 9. Comparative Financial Statistics.

part of their curriculum. Already a third of the membership of the Association are primary teachers and this must surely be set to grow.

Other avenues of National Curriculum guidance being developed are the Regional Inset seminars aimed at Inset providers. The Association hosted a series of ten regional seminars during the period November 1991 to March 1992. The purpose was to explore Inset needs related to the National Curriculum, to exchange views on current Inset provision and to identify opportunities for co-operation over supplying future needs. The seminars had the approval of DES, NCC and also LEAs and the Association believes that given access to funding a secure Inset framework can be established to ensure good quality geographical education for the National Curriculum.

The 1993 Centenary Conference is scheduled to take place, very appropriately, in Sheffield whilst members of the Association can also look forward to a variety of other events including a special conference and dinner in Oxford in July 1993, a meeting of the European Geographical Education Associations, and also a commemorative Channel Tunnel Excursion, if possible.

# Epilogue

The attainment of a hundred years in age is a noteworthy event in the life of any association, institution, society or individual: many are born but few survive this length of time. In the case of institutions it is inevitably a time for looking back, taking stock and then looking forward. So far we have concentrated upon the founding, growth and development of the Geographical Association and can now ask and answer the question of how far have the aims of our founders been achieved?

Bearing in mind that at the beginning of our century there was hardly any geography, as we now understand it, being taught in either the schools or universities in the British Isles, there can be little doubt that our founders would be greatly pleased with the present situation where geography has become a compulsory foundation subject in all maintained schools. The situation in the Universities however has become somewhat complicated as a result of the upgrading of most Polytechnics to University status during 1992–93. Nearly all of the pre-1992 Universities and some of the Polytechnics had by this date acquired Honours Schools in Geography but a number of the Polytechnics had developed Environmental Science or Earth Science Schools or Departments. These were very often staffed by geographers and the courses given were little more than geography under another name.

The Association can reasonably claim to have been largely responsible for bringing about the considerable improvement in geographical studies after the initial impetus given by a handful of dedicated geographers working under the patronage of the Royal Geographical Society. It has not been an easy road and problems still remain but steady progress has been made throughout the century: even in 1935 H. J. Mackinder was able to write to H. R. Mill saying 'It is indeed for both of us a glorious thing that we have lived to see what is the beginning of the triumph of our youthful ideas' (RGS archives). How enthusiastically would Mackinder have greeted the Association's centenary and how pleased our Founders would be to learn that HRH The Princess Royal, perhaps the most travelled member of the Royal Family, has graciously accepted an invitation to become Patron of the Association's Centenary celebrations.

Not only is geography now firmly anchored in the educational system of the country but we have reached the stage where its specialist products have permeated into all aspects of the economic, commercial, industrial, political, administrative and social life of the national community. A recent careers survey revealed over a hundred different professions where geography graduates were known to be employed. Geographers have now risen to be Secretary of State for Education. Vice-Chancellors of Universities, Director Generals of the Ordnance and Military Surveys, leading politicians and media personalities: they have also been much favoured as Headmasters and Headmistresses of Maintained, Public and Private schools. All those concerned with the natural, social and economic environment at home and overseas need geography. Its use by the sailor, soldier and airman is so obvious as to need no emphasis: it is equally true that no responsible person, journalist, business person or administrator who is involved with the real world can rightly pass judgement on problems concerning countries with whose geographical background he or she is not familiar.

It is sometimes stated that the possession of a geography degree is irrelevant to many of the posts to which geographers have been appointed. Such critics overlook the fact that a geographical training and the adaptability acquired appeals to many employers who have become aware of the fact that graduates in geography, unlike those in many other disciplines, have had a training in *each* of the four basic communication skills of literacy, numeracy, articulacy and graphicacy. Essays, seminar work, field work and projects build up an expertise in *Literacy*; laboratory work, statistical exercises and computer training underwrite an expertise in *Numeracy*; tutorials and field interviews with project presentation strengthen the skill of *Articulacy*; whilst unique to geography is the whole range of visual-spatial data presentation and analysis represented by maps, graphs, charts, diagrams and computer graphics collectively expressed by the term *Graphicacy*.

So far so good. But not all problems have been resolved. The subject did not achieve core status in the National Curriculum which was the reasoned hope of the profession, in the national curriculum discussion; the view that it should was also supported by many amongst the educated public (v. *The Times* editorial). The problem of Key Stage 4 has yet to be resolved and the effects have yet to be discovered. Furthermore will there be a Key Stage 5 for 16+ pupils? Additionally geography has to watch that its present cohesion is not destroyed by 'progressive' ideas about cross-curricular themes, multicultural courses, social studies, the humanities approach and sundry other popular notions that surface from time to time.

Clearly then there will be much work for the Geographical Association as it enters its second century: moreover it also has its own problems to face. Most notably there is the need to possess its own Headquarters, whilst the long term effect of the 1977 change in the Constitution is as yet unknown. Much depended in the early days and until quite recently on the dedicated voluntary service of a few university individuals acting over long periods. Will the future membership of the Association be able to provide a continual supply of enthusiastic volunteers to fill the short term appointments of what are now 28 unpaid 'officers' of the Association?* Furthermore the democratisation of election procedures combined with the short service periods has led to a serious diminution of the University/professorial element in the organisational structure. There are obvious dangers in this situation. The Association needs co-operation at all levels: a divorce between school and university approaches could be disastrous.

We cannot conclude this account without emphasising once again the way in which past and present members at both Headquarters and Branch levels have worked voluntarily for the good of their colleagues and the Association. Who better to pay this tribute than Professor Alice Garnett who wrote in her contribution to the 1964 IGU volume: 'When the history of our Association is written in

years to come there should be recorded in due perspective the great debt that the Association owes to the many distinguished contributors who have given their services, writing and preparing texts and pages for publication, without question of fee or royalty. The proceeds from publication sales have indeed played a major part in ensuring financial security for the Association during difficult times of inflationary changes and the social developments necessary in a welfare state. Gone are the days in Great Britain when voluntary organisations such as ours could find and depend on staff who all but gave their services. Even less than two decades ago our office work was conducted by a chief clerk employed at a salary of less than £200 per annum, without the security of even the most modest pension scheme. When now we take justifiable pride in the complete revolution in the administration of our affairs that conditions 'in the sixties' have properly demanded, it is fitting that we do not forget the labours of such persons, for they kept our Association in being during the hard conditions of the pre-war years'.

We now live in a complex society and a world of instant communication in which occurrences in one place often produce a chain of events which span the globe. We depend upon an increasingly fragile physical environment whose complex interactions require sophisticated analysis and sensitive management. The world now faces problems of population increase, environmental degradation, deforestation, climatic change, atmospheric pollution, ozone depletion, global warming, sea level variation, natural and man-made disaster situations, water shortage, sustainability and carrying capacity to name but a few of the more serious difficulties. To understand the complexities of these problems and all aspects of human life on Planet Earth, and in seeking solutions, requires a basic knowledge that only geography can give. There will thus be no shortage of tasks for the Association to turn to in the immediate future but it can enter its second century with a confidence based upon the solid achievements of its first century.

*The compiler of the Association's bi-centennial history will face a daunting task!

## Officers

| | 1991–92 | 1992–93 |
|---|---|---|
| President: | Eleanor Rawling | Simon Catling |
| Past President: | Bryan Coates | Eleanor Rawling |
| Senior Vice-President: | Simon Catling | Andrew Goudie |
| Junior Vice-President | Andrew Goudie | Tony Binns |
| Honorary Vice-President: | Doreen Massey | Doreen Massey |
| | Tony Thomas | Tony Thomas |
| Trustees: | Rex Walford | Neil Simmonds |
| | Patrick Bailey | Patrick Bailey |
| | Neil Simmonds | Rex Walford |
| Joint Honorary Secretaries: | Peter Fox | David Burtenshaw |
| | David Burtenshaw | Peter Fox |
| Hon. Assistant Secretaries: | Tony Pearce | Tony Pearce |
| | Margaret Roberts | Margaret Roberts |
| Honorary Treasurer: | Neil Simmonds | Neil Simmonds |
| Hon. Assistant Treasurer: | Bob May | Bob May |
| Chair, ESC | Michael Hewitt | Michael Hewitt |
| Chair, P&CSC | Geoff Sherlock | Judith Mansell |
| Publications Officer: | James Hindson | James Hindson |
| Library/Information Officer: | Kevin Luker | Kevin Luker |
| Branch Officer: | Bob Jones | Bob Jones |
| Publicity Officer: | Keith Hilton | Keith Hilton |
| Annual Conference Officer: | Russell Chapman | Russell Chapman |
| Asst. Ann. Conf. Officer: | Ralph Dunkley | Ralph Dunkley |
| Hon. Editor: *Geography* | Derek Spooner | Derek Spooner |
| *Teaching Geography* | David Boardman | David Boardman |
| *Primary Geographer* | Wendy Morgan | Wendy Morgan |
| *GA News* | Bob Leake | Bob Leake |

## Headquarters Staff 1993

| | |
|---|---|
| Senior Administrator: | Frances Soar |
| Administrator: | Julia Legg |
| Assistant Editor: | Noreen Pleavin |
| Administrative Assistant: | Barbara Spooner |
| Editorial Assistant: | Diane Wright |
| Publications Secretary: | Susan Moat |
| Administrative Secretary: | Margaret Vickers |
| Administrative Secretary: | Ann Armstrong |
| Accounts Clerk: | Sue Mappin |
| Clerical Assistant: | Christina Leybourne |
| Clerical Assistant: | Brenda Porter |
| Clerical Trainee: | Nicola Wadsworth |

Figure 10. Officers of the Association and Headquarters Staff.

# Appendix A

## The Branch Structure of the Association

No historical account of the Geographical Association would be complete without details of the evolution and work of its branches. As noted in the main text branches began to appear naturally in the very early days as geography teachers foregathered to discuss their problems and exchange information. The first to emerge was in South London on 17th May 1904, quickly followed by others in Bedfordshire 13th May 1905, Bournemouth 17th May 1905, Sheffield 26th February 1907, Bristol 22nd March 1907, Huddersfield 1907, North London 1907, Manchester 1909, Chester 1911, and Leeds 1911. By 1914 Southampton and Birmingham had also been added and a countrywide pattern was clearly appearing. This was a spontaneous reaction in support of the national movement being sponsored by the Association.

Happily World War 1 did not decimate the work of the branches as did World War 2 and as soon as hostilities ceased in 1918 the way was open for a rapid spread of the branch network from the existing base. By 1920 Central Lancashire (Preston and Blackpool), North Lancashire, Plymouth, Malvern, Wigan, Cardiff, North Wales (Bangor), Bishop Auckland, Crewe, East Suffolk, Essex, Exeter, Hereford, Lincoln, Leicester, Nottingham, Swansea, Thanet, Tottenham and York all had branches and the foundations were laid for a branch structure which was to persist for the rest of the century.

The overall pattern however has not been static: apart from the larger urban centres branches have appeared, disappeared, and reappeared with remarkable regularity as so much has depended on voluntary officers prepared to sacrifice time, energy and often their own finance in the furtherance of the subject. Promotions, retirements and deaths of key personnel have often led to the temporary demise of a branch until another volunteer has emerged to resuscitate a group. Altogether 150 locations of branches in the British Isles have been recorded but at any one time the maxi-mum active has never exceeded 75: these, however, have ensured continuous participation at the grass roots as in many cases members have been able to transfer from a lapsed branch to an active one.

Part of the changing pattern is also related to the emergence of separate organisations in Ireland and Scotland.

Each branch is unique in the approach to its organisation and subscription, as the only constitutional requirement of the Geographical Association is that at least two Branch Officers and four others must be full members of the Association and that each September annual reports of finance, membership and programmes must be submitted to Headquarters. The two Branch Officers are the key to the success or failure of a branch—success depends on a minimum of two individuals with initiative being prepared to share the duties of Chairman/President and Secretary/Treasurer: and ideally these should be supported by an active and interested committee. Because potential personnel are limited in smaller urban centres this situation has led to some remarkable periods of long term service in the cause of the subject: some examples are noted below.

It is from the branches of course that so many of the national officers and committee members are recruited and there is a natural progression from branch committee work to national committee service and thence to higher (and more onerous) office on Council. At least half the recent members of Council have been active branch members. Both the Swansea and Bristol branches are proud of the fact that they have each produced three national presidents in recent years. Membership of the branches has varied from the proverbial handful to over a thousand in the case of Liverpool and Birmingham with the latter holding the record of 1,250 in 1972. The bulk of the branch membership of course falls into the student or associate category and council would like to see more of the branch

# Locations of the Association's Branches 1893–1993

Ambleside (Lake District)
Ashford
Aylesbury
Banbury
Bangor (North Wales)
Barnstable
Bedford (shire)
Belfast
Berkhamsted
Bexleyheath (NW. Kent)
Birmingham
Bishop Auckland
Blackburn
Blackpool
Blandford (Dorset)
Bognor
Bolton
Bournemouth
Bradford
Braintree (Essex)
Bridgend
Bridgenorth (Shropshire)
Brighton
Bristol
Bromley (Kent)
Cambridge
Camelford (N. Cornwall)
Canterbury
Cardiff
Carlisle
Carmarthen
Chelmsford (W. Essex)
Chester
Chorley (Lancashire)
Colchester (Essex)
Coleraine (N. Ireland)
Coventry
Crewe
Croydon
Dartford
Derby
Doncaster
Dorchester (S. Dorset)
Dorking
Dunfermline
Durham
Eastbourne

Edinburgh
Enfield (E. Middlesex)
Essex SE
Essex SW
Exeter
Furness
Glasgow
Gloucester
Grimsby
Guildford
Hastings
Helston (Cornwall)
Hereford
Huddersfield
Hull
Ipswich
Isle of Man
Isle of Thanet (Kent)
Isle of Wight
Kingston upon Thames
Lancaster
Leeds
Leicester
Lincoln
Liverpool
Londonderry
London Central
London E. Middlesex
London Goldsmiths
London Harrow
London Isleworth
London KCL/LSE
London North
London South
London South West
London Tottenham
London West Ham
Luton
Maidstone
Malvern
Manchester
Mansfield
Medway
Newcastle upon Tyne
Newry (Ireland)
Nottingham
Northampton

Norwich
Oldham
Ongar (Essex)
Oxford
Penzance
Peterborough
Plymouth
Portsmouth
Preston
Reading
Romford (SW. Essex)
St. Andrews
St. Austell
St. Ives (Cornwall)
Scarborough
Sheffield
Shrewsbury
Sidcup (Kent)
Slough
Southampton
Stafford
Stockton on Tees
Stoke on Trent
Suffolk
Sunderland
Sussex (West)
Swansea
Swindon
Taunton
Torquay
Truro
Tunbridge Wells
Tyneside
Walsall
Waltham Forest
Warrington
Watford
West Bromwich
Weymouth
Wickford (Essex)
Wigan
Winchester
Wolverhampton
Worcester
Worthing
Wrexham
York

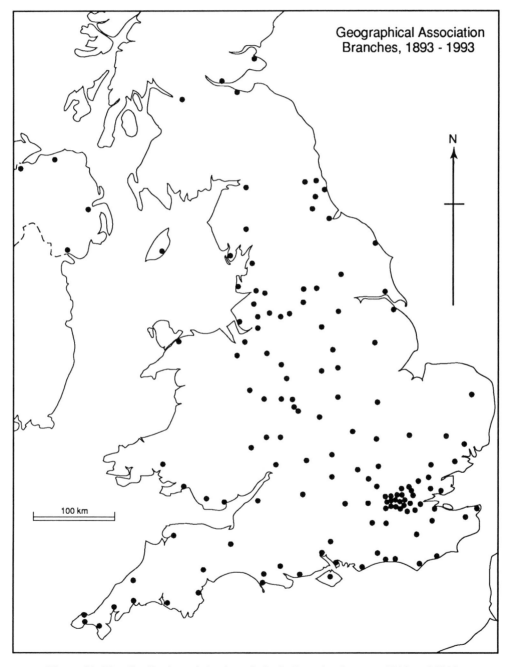

Figure 11. The distribution of the Association's Branches between 1893 and 1993.

# Location of Spring Conferences 1921–1974

| | | | | |
|---|---|---|---|---|
| 1921 | Southampton | | 1948 | Birmingham |
| 1922 | Leeds | | 1949 | Liverpool |
| 1923 | Shrewsbury | | 1950 | Falmouth |
| 1924 | Exeter | | 1951 | Hull |
| 1925 | Reading | | 1952 | Tenby |
| 1926 | Bristol | | 1953 | Lincoln |
| 1927 | Liverpool | | 1954 | Exeter |
| 1928 | Oxford | | 1955 | York |
| 1929 | Hull | | 1956 | Brighton |
| 1930 | Birmingham | | 1957 | Matlock |
| 1931 | Manchester | | 1958 | Aberystwyth |
| 1932 | London | | 1959 | Leicester |
| 1933 | Liverpool | | 1960 | Durham |
| 1934 | Glasgow | | 1961 | Bristol |
| 1935 | Nottingham | | 1962 | Keele |
| 1936 | Sheffield | | 1963 | Swansea |
| 1937 | Swansea | | 1964 | Falmouth |
| 1938 | Durham | | 1965 | Birmingham |
| 1939 | Leicester | | 1966 | Oxford |
| 1940 | Blackpool | | 1967 | Sheffield |
| 1941 | Edinburgh (Combined Annual/Spring) | | 1968 | Nottingham |
| 1942 | Exeter (Combined Annual/Spring) | | 1969 | Brighton |
| 1943 | Cambridge (Combined Annual/Spring/Summer) | | 1970 | Liverpool |
| 1944 | – – – | | 1971 | – – – (postal strike) |
| 1945 | – – – | | 1972 | Newcastle upon Tyne |
| 1946 | Caernarvon | | 1973 | Southampton |
| 1947 | Sheffield | | 1974 | Worcester |

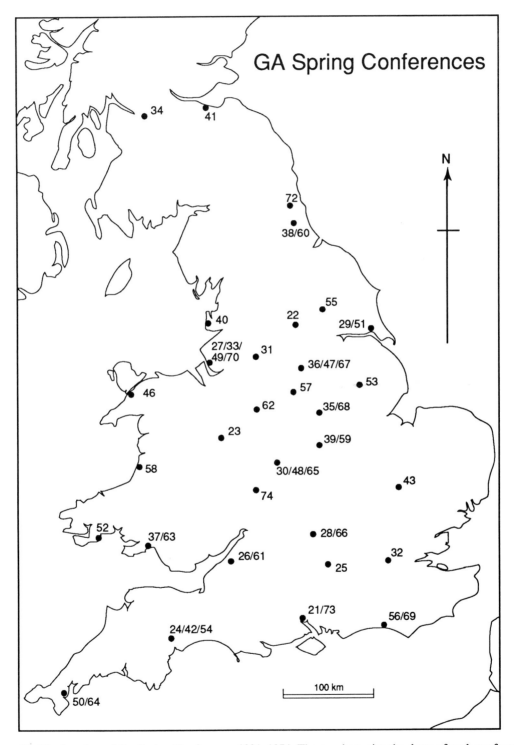

Figure 12. The location of the Spring Conferences 1921–1974. The numbers give the dates of each conference.

membership up-graded to full membership of the Association.

An academic lecture programme is the backbone of branch activity and here the Association acts as a catalyst bringing in voluntary lecturers from universities, polytechnics and local colleges. Whilst many lectures are of general interest there has been a growing tendency to cater more especially for sixth form audiences and to include topics of relevance to Advanced Level syllabuses. The needs of sixth formers are also catered for by means of Sixth Form Conferences (originally introduced by the Bolton Branch) and meetings devoted to career possibilities.

Field excursions also figure in many programmes and Association branches have a long history of activity in this respect: pioneer work was undertaken well in advance of the package tour industry and the spread of the motor car—both of which have, more recently, blunted the popularity of both the long and short term organised expedition. Local field trips and excursions have however been a regular feature on most branch programmes throughout the century and these have often led to more spectacular ventures such as the Lupton continental tours of the 1920s; the famous day visit of 890 pupils to York and Whitby organised by the Bradford Branch in 1933; the Mersey sailings mounted by the Liverpool Branch; the remarkable inter-continental study tours of the 1960s and 1970s organised by the Isle of Thanet Branch to the USA, Canada, West Indies and East Africa; and more recently European visits by the Gloucester and Cheltenham Branch.

Another branch interest has been involvement with publications, usually of a local character. Coventry, Cambridge, Blackpool and and the Isle of Thanet have been notable in this respect with more recent contributions from Birmingham, Norfolk, Lincoln and Tyneside; whilst Southampton has produced filmstrips and NW Ulster teaching kits. Additionally both Sheffield and Birmingham have sponsored national Association publications from their branch funds.

The branch structure has also been of considerable value from time to time when national exercises have surfaced: particular examples include the First (Dudley Stamp) Land Utilisation Survey of the 1930s; the Second (Alice Coleman) Land Utilisation Survey of the 1960s; the Map Watch of the 1980s; the World Wise Quiz competition of the 1980s and 1990s; and the curriculum debate of the 1980s. The branches have played a major role in all these activities.

One important historic function of the branches, now alas no longer possible, was to host the mobile Spring Conference which was held annually, apart from the war years, between 1921 and 1974. The Spring Conference was another natural growth which began at Southampton in 1921 by invitation of the Portsmouth, Southampton and Bournemouth branches. The success of the meeting owed much to the top level support given by Sir Charles Close, then Director-General of the Ordnance Survey, and Sir Halford Mackinder, then at the height of his career: it led to an annual spring event migrating around the regions and hosted by the branches. Eventually attracting between 2–300 participants each year the preferred locations for the Spring Conferences became the university towns as these could provide modest priced accommodation.

World War 2 produced a temporary break in the sequence of spring conferences but this popular event in the annual life cycle of the Association was quickly resumed after 1945. Sadly however its demise came in the 1970s when the Heath Government declared January 1st a public holiday as from 1974. This forced the Association to move the Annual Conference from its traditional time in the Christmas vacation to the Easter vacation and the migratory Spring Conference ceased thereafter (see Fig. 12).

Finally many branches have a social aspect: annual dinners, annual socials, overseas excursions (Paris, Netherlands, etc.) and river trips (Thames, Cherwell and Mersey) where the contents are as much social as geographical, have figured in many branch programmes.

The limited number of adult members in many branches has inevitably meant that the chores of running a branch have had to be borne by the same individuals over lengthy periods of time. It is a sign of the strength and dedication of the officers and the hold of the subject that the branch fabric has, like the national organisation, survived the century with such success, despite being dependent on voluntary workers. Unsung heroes abound in the fragmented branch records and mention can only be made of a few exceptional cases. Foremost amongst these must be the late Professor Estyn Evans' record of 60 years' continuous service as Hon. Secretary, President and Vice-President of the Belfast branch between 1929 and 1989. Professor J. K. Charlesworth was also in continuous office as President or Vice-President of the Belfast Branch for 43 years from 1929 to 1972, whilst the Hon. Treasurer, R. Lyttle, served for 41 years from 1944 to 1985.

Other more recent long serving officers include Mr. Edmund Dobson, a founder-member of the Winchester Branch, who retired in 1991 after acting as Chairman for 43 years. Still in office is Mr. Anthony R. Wheeler who was Secretary of the Worcester Branch from 1951 to 1987 when he became and still continues as its Branch President—a total of 42 years' continuous service. Also still in office as President of the Isle of Thanet Branch is Professor Alice Coleman with 35 years of service to her credit and all the more remarkable in this case as she has delivered an annual presidential address each year to the branch on 35 different topics. She is closely followed by Miss Dorothy Sylvester's record of 33 years' presidency of the Crewe and Nantwich Branch.

Other notable periods of long service include Miss Rosemary Robson who has been Programme Secretary of the Eastbourne Branch for 35 years (1956 to date): Miss Marjorie Woodward who has been Hon. Secretary, Tours Secretary and Chairman of the Isle of Thanet Branch for 35 years (1956 to date). Two long periods of service in the Banbury Branch are Mr. Brian Little as Hon.

Secretary for 28 years (1963 to date) and Mr. John Gilchrist, Hon. Treasurer for 25 years (1966 to date). Mr. Malcolm Eames was Hon. Treasurer of Berkhamsted for 35 years (1952–1987). The Blackpool Branch claims a considerable number of long serving officers from 1952 onwards: The President Mr. A. J. Clarke, the Hon Auditor Mr. J. S. Darwell both record more than 25 years, whilst Mrs. I. Darwell, Miss Ellwood, Mrs. M. Gaskell, Miss L. M. Moore and Miss M. Muschamp also have long records in various offices. Professor W. G. V. Balchin, President of the Swansea Branch for 24 years (1954–78), can perhaps lay claim to another record as he has recently been reincarnated as President of the Bradford Branch in succession to the late Professor Kenneth Baker, who recently completed 21 years of service (1969–90).

There are doubtless many other branch members from the early part of the century who have completed long service periods but few branches have continuous records as a result of changing personnel and the two world war breaks. To all these unsung heroes, however, the profession, the Association, and the educational world, owe a debt of gratitude for the hours of unpaid voluntary work so readily given.

The importance of the Branch network to the Association was formally recognised in the revised 1977 constitution by the creation of a Branch Officer: this post was filled initially by Miss Sheila Jones (1978–83), then by Robert May (1983–89) and Robert G. Jones (1989–   ). The Branch Officer is supported by a Newsletter circulated twice a year to all branches. The location of active Branches of the Association as the centenary approaches is given in Figure 13 whilst the names and addresses of the Secretaries in each case are given on pages 91 and 92.

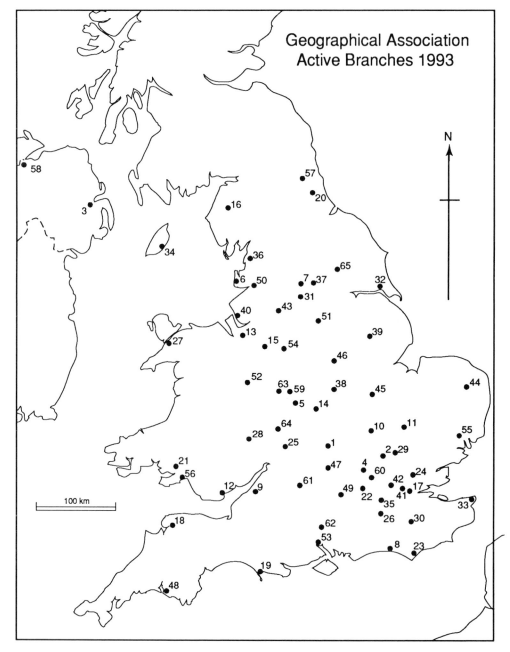

Figure 13. Active Branches of the Association 1992–1993. The numbers locate the Branches in
the following lists, pp. 91 and 92.

# Secretaries of GA Branches 1992–93

This list was correct at the time of writing. However since then changes will have occurred.

| | | |
|---|---|---|
| **Branch Officer:** | | Robert G. Jones, 29 Granville Terrace, Stone, Staffs. ST15 8DF. Tel. 0785 815521 |
| 1. | Banbury | Mr. B. Little, 12 Longfellow Road, Banbury, Oxfordshire OX16 9LB. |
| 2. | Bedfordshire | Mr. A. D. Cooper, Luton College of HE, Park Square, Luton, Bedfordshire LU1 3JU. |
| 3. | Belfast | Mrs. J. Neill, 17 Kirkwood Park, Saintfield, Belfast, Co. Down BT24 7DP. |
| 4. | Berkhamsted | Mrs. L. Waumsley, Kingsley Orchard, Chesham Road, Wiggington, Near Tring, Herts. HP23 6HN. |
| 5. | Birmingham | Mrs. S. M. Whitehand, The Gables, 12 Fiery Hill Road, Barnt Green, Worcestershire B45 8LG. |
| 6. | Blackpool | Miss P. Dowson, 63 Harrington Avenue, Blackpool, Lancashire FY4 1QE. |
| 7. | Bradford | Dr. D. Cotton, School of Environmental Sciences, University of Bradford, Bradford, West Yorkshire BD7 1DP. |
| 8. | Brighton and District | Ms. M. Zakiewicz, 50 Withdean Crescent, Brighton, Sussex BN1 6WH. |
| 9. | Bristol | Mrs Michelle Dimes, 16 Hawthorn Close, Charfield, Wotton-under-Edge, Gloucestershire GL12 8TX. |
| 10. | Buckinghamshire | Mrs. M. Murray, c/o Ousedale School, The Grove, Newport Pagnell MK16 0BJ. |
| 11. | Cambridge | Dr. H. Allen, Homerton College, University of Cambridge, Hills Road, Cambridge CB2 2PH. |
| 12. | Cardiff and South East Wales | Mr. J. E. G. Worth, 42 Redbrook Road, Newport, Gwent NP9 5AB. |
| 13. | Chester | Mrs. J. Evans, 14 Guilden Green, Guilden Sutton, Chester CH3 7SP. |
| 14. | Coventry | Mr. Brian Ellis, Dept. of Science Education, University of Warwick, Coventry, West Midlands CV4 7AL. |
| 15. | Crewe and Nantwich | Mr. D. A. Atkinson, 37 Sandringham Drive, Wistaston, Crewe, Cheshire CW2 8JF. |
| 16. | Cumbria | Mrs. Rachael Tickner, 26 Mayo Street, Cockermouth, Cumbria CA13 0BY. |
| 17. | London Central | Dr. A. Warren, Dept. of Geography, University College, 26 Bedford Way, London WC1H 0AP. |
| 18. | Devon (North) | Mr. J. Lifford, North Devon College, Barnstaple, North Devon EX31 2BQ. |
| 19. | Dorset | Mr. A. Holiday, Weymouth College, Cranford Avenue, Weymouth, Dorset DT4 7LQ. |
| 20. | Durham | Dr. E. W. Anderson, Department of Geography, University of Durham, South Rd., Durham, Co. Durham DH1 3LE. |
| 21. | Dyfed | Mr. A. Carter, 54 Golwg-yr-Mynydd, Fforest Pontardulais, West Glamorgan. |
| 22. | Ealing and West Middlesex | Mr. B. Harris, Ealing College of Higher Education, St. Mary's Road, London W5 5RF. |
| 23. | Eastbourne | Mrs. B. A. Hinton, 11 Milnthorpe Road, Eastbourne, East Sussex BN20 7NS. |
| 24. | Essex | Ms. E. Rose, Meadgate Curriculum Development Centre, Mascalls Way, Meadgate Ave., Great Baddow, Chelmsford, Essex CM2 7NS. |
| 25. | Gloucester and Cheltenham | Mr. F. Harris, College of St. Paul and St. Mary, The Park, Cheltenham, Gloucestershire GL50 5SR. |
| 26. | Guildford | Mr. R. S. Burns, Earlsdon, 18 Ellis Avenue, Onslow Village, Guildford, Surrey GU2 5SR. |

| 27. | Gwynedd | Mr. John Williams, Head of Dept. of Geography, Ysgol Sir Hugh Owen, Caernarfon, Gwynedd. |
| 28. | Hereford | Mrs. L. Miles, Netherton House, Harewood End, Hereford, Herefordshire HR2 8LA. |
| 29. | Hertfordshire Geography Teachers Association | Mr. J. Burden, SUSC, Pin Green School Site, Lonsdale Road, Stevenage, Hertfordshire SG1 5DQ |
| 30. | High Weald | Miss Joan Lowdon, 8 Park Road, Tunbridge Wells, Kent TN4 9JN. |
| 31. | Huddersfield and Halifax | Mr. C. D. Holmes, 80 Hill Top Road, Selendine Nook, Huddersfield, West Yorkshire HD3 3SJ. |
| 32. | Hull | Mr. M. Wilson, 72 Summergangs Road, Hull, North Humberside HU8 8LP. |
| 33. | Isle of Thanet | Mrs. E. M. Buller, 26 Saltwood Gardens, Cliftonville, Margate, Kent CT9 3HQ. |
| 34. | Isle of Man | Mr. John Cain, Isle of Man Education Dept., 1 Government Buildings, Bucks Road, Douglas, Isle of Man. |
| 35. | Kingston upon Thames | Dr. N. Walford, School of Geography, Kingston Polytechnic, Penrhyn Road, Kingston upon Thames, Surrey KT1 2EE. |
| 36. | Lancaster | Mr. G. Bolton, Geography Department, St. Martins College, Bowerham, Lancaster LA1 3JD. |
| 37. | Leeds | Mr. R. Jackson, 53 Westcombe Avenue, Roundhay, Leeds, West Yorkshire LS8 2BS. |
| 38. | Leicester | Mr. A. J. Budd, Geography Department, University of Leicester, University Road, Leicester LE1 7RH. |
| 39. | Lincoln | Mr. A. G. Bilham-Boult, Freiston Hall Field Centre, Haltoft End, Freiston, Boston, Lincolnshire PE22 0NX. |
| 40. | Liverpool | Mr. G. Baker, 46 Oxford Drive, Waterloo, Merseyside L22 7RZ. |
| 41. | London (North East) | Ms. I. Caplin, Highams Park Senior High Sch., Handsworth Avenue, London E4 9PJ. |
| 42. | London (North) | Mr. R. Knowles, Department of Geography, Polytechnic of North London, Marlboro' Bldg., 383 Holloway Rd, London N7 6PN. |
| 43. | Manchester | Mrs. J. Sawyer, 1 Gorsey Road, Wilmslow, Cheshire SK9 5DU. |
| 44. | Norfolk | Mrs. E. Plater, 66 Damgate Street, Wymondham, Norfolk NR18 0BH. |
| 45. | Northamptonshire | Miss Fiona Parkinson, 1 Meadow Close, Little Cransley, Kettering, Northamptonshire NN14. |
| 46. | Nottingham | Mrs. E. M. Garnham, 121 Selby Road, West Bridgford, Nottingham, Nottinghamshire NG2 7BB. |
| 47. | Oxford | Ms. N. Lindsay, The Cherwell School, Marston Ferry Road, Oxford, Oxfordshire OX2 7EE. |
| 48. | Plymouth and District | Mrs. L. B. Anderson, 56 Colesdown Hill, Plymouth, Devon PL9 8AS. |
| 49. | Reading and District | Mrs. Angela Holden, Kendrick School, London Road, Reading, Berkshire RG1 5BN. |
| 50. | Ribblesdale | Mr. and Mrs. M. Pearson, Carrwood House, Hothersall Lane, Hothersall, Preston, Lancashire PR3 2XB. |
| 51. | Sheffield and District | Mr. Kevin Luker, 226 Carterknowle Road, Sheffield, South Yorkshire S7 2EB. |
| 52. | Shropshire | Dr. J. C. Hindson, 16 Cedar Close, Bayston Hill, Shrewsbury SY3 0PD. |
| 53. | Southampton | Mrs. K. Adams, 55 Springvale Road, Kings Worthy, Winchester, Hampshire SO23 7ND. |
| 54. | Staffordshire (North) | Mr. R. G. Jones, 29 Granville Terrace, Stone, Staffordshire ST15 8DF. |
| 55. | Suffolk | Mr. J. Hughes, 54 Upland Road, Ipswich, Suffolk IP4 5BT. |
| 56. | Swansea and District | Ms. A. Jones, 10 New Road, Cockett, Swansea, West Glamorgan SA2 0GA. |
| 57. | Tyneside | Mrs. D. E. Ward, Washington School, Spout Lane, Washington, Tyne and Wear NE37 2AA. |
| 58. | Ulster (North West) | Mrs. E. Ward, 27 Prospect Road, Portstewart, Co. Londonderry BT55 7NF. |

| | | |
|---|---|---|
| 59. | Walsall | The Geographical Association, Walsall Branch, c/o 343 Fulwood Road, Sheffield S10 3BP. |
| 60. | Watford | Mr. D. Trebble, 5 Askew Road, Moor Park, Northwood, Middlesex HA6 2JE. |
| 61. | Wiltshire | Mr. M. Turnbull, The North Wiltshire Centre, Drove Road, Swindon, Wiltshire SN1 3QQ. |
| 62. | Winchester and District | Ms. J. Brennan, 88 Teg Down Meads, Winchester, Hampshire SO22 5ND. |
| 63. | Wolverhampton | Mrs. A. Shortland, 5 The Ring, Little Haywood, Stafford ST18 0TP. |
| 64. | Worcester | Mr. C. D. A. Carter, 19 Whitewood Way, Worcester WR5 2LN. |
| 65. | York | Miss H. Arnold, 14 St James' Mount, York, YO2 1EL. |

# Appendix B
## Membership Numbers of the Association

Fluctuations in the growth of the membership are related to the effect of two World Wars, necessary increases in subscription levels as a result of inflationary trends, variations in the birth rate and curriculum changes. These factors have operated at different times and are dealt with in the main text.

| Year | No. | Year | No. | Year | No. | Year | No. |
|------|------|------|------|------|------|------|------|
| 1893 | 35 | 1918 | 1,458 | 1943 | 2,464 | 1968 | 8,522 |
| 1894 | 50 | 1919 | 2,379 | 1944 | 2,747 | 1969 | 9,138 |
| 1895 | 60 | 1920 | 3,695 | 1945 | 3,191 | 1970 | 7,991 |
| 1896 | 72 | 1921 | 4,159 | 1946 | 4,265 | 1971 | 8,387 |
| 1897 | 73 | 1922 | 4,462 | 1947 | 5,200 | 1972 | 8,301 |
| 1898 | 98 | 1923 | 4,510 | 1948 | 4,400 | 1973 | 8,400 |
| 1899 | 105 | 1924 | 4,610 | 1949 | 4,051 | 1974 | 8,522 |
| 1900 | 121 | 1925 | 4,585 | 1950 | 2,944 | 1975 | 8,470 |
| 1901 | 202 | 1926 | 4,351 | 1951 | 2,994 | 1976 | 7,941 |
| 1902 | 287 | 1927 | 4,447 | 1952 | 3,092 | 1977 | 7,963 |
| 1903 | 341 | 1928 | 4,449 | 1953 | 3,383 | 1978 | 8,049 |
| 1904 | 448 | 1929 | 4,345 | 1954 | 3,738 | 1979 | 8,024 |
| 1905 | 503 | 1930 | 4,233 | 1955 | 3,781 | 1980 | 7,203 |
| 1906 | 535 | 1931 | 4,293 | 1956 | 4,109 | 1981 | 6,635 |
| 1907 | 643 | 1932 | 4,009 | 1957 | 4,366 | 1982 | 6,165 |
| 1908 | 793 | 1933 | 4,036 | 1958 | 4,610 | 1983 | 6,258 |
| 1909 | 884 | 1934 | 4,002 | 1959 | 4,919 | 1984 | 6,463 |
| 1910 | 902 | 1935 | 4,001 | 1960 | 5,110 | 1985 | 6,555 |
| 1911 | 962 | 1936 | 3,867 | 1961 | 5,349 | 1986 | 6,520 |
| 1912 | 1,000 | 1937 | 3,666 | 1962 | 5,391 | 1987 | 7,191 |
| 1913 | 1,076 | 1938 | 3,646 | 1963 | 6,177 | 1988 | 7,044 |
| 1914 | 1,144 | 1939 | 3,384 | 1964 | 6,704 | 1989 | 6,873 |
| 1915 | 1,107 | 1940 | 2,554 | 1965 | 7,357 | 1990 | 8,499 |
| 1916 | 1,002 | 1941 | 2,261 | 1966 | 7,883 | 1991 | 9,672 |
| 1917 | 996 | 1942 | 2,305 | 1967 | 8,448 | 1992 | 10,391 |

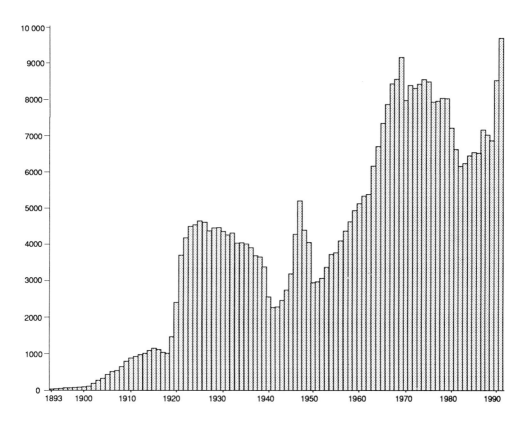

Figure 14. Fluctuations in the growth of membership of the Association.

# Appendix C

## Presidents of the Association

| | |
|---|---|
| 1897–1911 | Douglas W. Freshfield |
| 1912 | Dr. G. R. Parkin |
| 1913 | Professor E. J. Garwood |
| 1914 | Dr. J. Scott Keltie (later Sir) |
| 1915 | Hilaire Belloc |
| 1916 | Halford J. Mackinder (later Sir) |
| 1917 | Colonel Sir Thomas H. Holdich, KCMG KCIE CB |
| 1918 | Sir William M. Ramsay |
| 1919 | Professor Grenville A. J. Cole, FRS |
| 1920 | Sir Charles P. Lucas, KCMG KCB |
| 1921 | Professor Gilbert Murray, LL.D. FBA |
| 1922 | Rt. Hon. Lord Robert Cecil, PC KC |
| 1923 | Sir John Russell FRS |
| 1924 | Professor Sir Richard A. Gregory |
| 1925 | Professor John L. Myres, CBE FBA (later Sir) |
| 1926 | Rt. Hon. W. G. A. Ormsby Gore (later Lord Harlech) |
| 1927 | Colonel Sir Charles Close, KBE CB CMG FRS (later Arden-Close) |
| 1928 | Dr. Vaughan Cornish |
| 1929 | Colonel Sir Henry G. Lyons, FRS |
| 1930 | B. B. Dickinson |
| 1931 | Sir Leslie Mackenzie |
| 1932 | Dr. Hugh Robert Mill |
| 1933 | Professor Percy Maude Roxby |
| 1934 | Rt. Hon. Lord Meston |
| 1935 | James Fairgrieve |
| 1936 | Sir Josiah Stamp GCB GBE FBA (later Lord Stamp) |
| 1937 | Professor Patrick Abercrombie (later Sir) |
| 1938 | Sir Thomas H. Holland, KCSI KCIE FRS |
| 1939–1941 | C. C. Carter |
| 1942–1945 | T. C. Warrington |
| 1946 | Sir Cyril Norwood |
| 1947 | Sir Alexander Carr-Saunders |
| 1948 | Professor H. J. Fleure, FRS |
| 1949 | Sir Harry Lindsay |
| 1950 | Professor L. Dudley Stamp (later Sir) |
| 1951 | Leonard Brooks |

| | |
|---|---|
| 1952 | Professor F. Debenham, OBE |
| 1953 | Dr. O. J. R. Howarth OBE |
| 1954 | Professor S. W. Wooldridge, CBE FRS |
| 1955 | L. S. Suggate |
| 1956 | Rt. Hon. Lord Nathan, PC TD |
| 1957 | Professor P. W. Bryan |
| 1958 | Professor R. Ogilvie Buchanan |
| 1959 | Professor J. A. Steers, CBE |
| 1960 | Professor A. Austin Miller |
| 1961 | Geoffrey E. Hutchings |
| 1962 | Professor E. G. Bowen |
| 1963 | Professor K. C. Edwards, CBE |
| 1964 | Professor D. L. Linton |
| 1965 | J. A. Morris |
| 1966 | Professor S. H. Beaver |
| 1967 | E. C. Marchant |
| 1968 | Professor Alice Garnett |
| 1969 | Professor J. R. James |
| 1970 | Mrs. I. M. Long |
| 1971 | Professor W. G. V. Balchin |
| 1972 | Alan D. Nicholls |
| 1973 | Professor R. W. Steel |
| 1974 | Professor H. Thorpe |
| 1975–1976 | Miss S. M. Jones |
| 1976–1977 | Professor M. J. Wise |
| 1977–1978 | Professor S. Gregory |
| 1978–1979 | Dr. N. J. Graves |
| 1979–1980 | Professor J. A. Patmore |
| 1980–1981 | V. D. Dennison |
| 1981–1982 | Professor W. R. Mead |
| 1982–1983 | Professor R. Lawton |
| 1983–1984 | R. A. Walford |
| 1984–1985 | Miss P. J. Cleverley |
| 1985–1986 | P. J. M. Bailey |
| 1986–1987 | Professor Denys Brunsden |
| 1987–1988 | Dr. G. Humphrys |
| 1988–1989 | M. J. Storm |
| 1989–1990 | R. A. Daugherty |
| 1990–1991 | B. E. Coates |
| 1991–1992 | Mrs. E. M. Rawling |
| 1992–1993 | S. Catling |
| 1993–1994 | Professor A. S. Goudie |

# Appendix D

## Honorary Secretaries of the Association

| | | | |
|---|---|---|---|
| 1893–1900 | B. B. Dickinson | 1976–1981 | R. A. Daugherty |
| 1900–1915 | Professor A. J. Herbertson | 1979–1984 | B. E. Coates |
| 1915–1917 | Office vacant | 1981–1985 | M. T. Williams |
| 1917–1946 | Professor H. J. Fleure | 1984–1987 | Elspeth M. Fyfe |
| 1947–1967 | Professor Alice Garnett | 1985–1989 | J. A. Binns |
| 1968–1973 | Professor S. Gregory | 1987–1993 | P. Fox |
| 1968–1976 | W. R. A. Ellis | 1989– | D. Burtenshaw |
| 1973–1979 | G. M. Lewis | | |

# Appendix E

## Honorary Editors of the Association

*The Geographical Teacher* and then *Geography*

| | | | |
|---|---|---|---|
| 1901–1902 | A. W. Andrews and A. J. Herbertson | 1983–1989 | Professor R. King |
| 1903–1915 | Professor A. J. Herbertson | 1990– | D. J. Spooner |
| 1916–1917 | H. O. Beckit and Professor P. M. Roxby | | |
| 1918–1946 | Professor H. J. Fleure with Professor P. M. Roxby as Associate Editor 1918–1932. Title changed to *Geography* in 1927. | | |

*Teaching Geography*

| | |
|---|---|
| 1975–1985 | P. J. M. Bailey |
| 1985–1988 | Mrs. E. M. Rawling |
| 1988–1990 | Mrs. J. Kelly |
| 1990– | D. Boardman |

| | |
|---|---|
| 1947–1964 | Professor D. L. Linton |
| 1965–1979 | Professor N. Pye |
| 1980–1983 | Professor R. J. Small |

*Primary Geographer*

| | |
|---|---|
| 1989– | Mrs. W. Morgan |

# Appendix F

## Honorary Treasurers of the Association

| | | | |
|---|---|---|---|
| 1893–1900 | C. E. B. Hewitt | 1966–1976 | Professor M. J. Wise |
| 1900–1907 | J. S. Masterman | 1976–1981 | D. G. Mills |
| 1907–1925 | E. F. Elton | 1981–1984 | F. E. I. Hamilton |
| 1926–1928 | Colonel Sir H. G. Lyons | 1984–1987 | B. E. Coates |
| 1931–1955 | Sir William Himbury | 1987–1988 | P. J. M. Bailey |
| 1955–1966 | Professor Sir L. Dudley Stamp | 1988– | N. M. Simmonds |

# Appendix G
## Trustees of the Association

| | | | |
|---|---|---|---|
| 1914–1917 and 1919–1947 | Sir Halford J. Mackinder | 1955–1973 | Professor E. G. Bowen |
| | | 1973–1979 and 1986–1990 | Professor S. Gregory |
| 1914–1947 | Professor L. W. Lyde | 1976–1981 | D. G. Mills |
| 1914–1922 | Rev. Dr. James Gow | 1977–1986 | J. Old |
| 1918–1919 | A. D. Carlyle | 1979–1988 | Professor J. A. Patmore |
| 1922–1926 | Principal J. H. Davies | 1984–1987 | B. E. Coates |
| 1927–1954 | Professor Sir J. L. Myres | 1987– | P. J. M. Bailey |
| 1947–1953 | Lord Rennell | 1988– | N. M. Simmonds |
| 1947–1964 | Leonard Brooks | 1990– | R. A. Walford |
| 1954–1966 | Professor Sir L. D. Stamp | | |
| 1955–1977 | Professor W. G. V. Balchin | | |

# Appendix H

## Honorary Conference Organisers of the Association

| | | | |
|---|---|---|---|
| 1922–1932 | V. C. Spary | 1969–1974 | D. Brunsden |
| 1933–1934 | H. J. Wood | 1975–1978 | D. K. C. Jones |
| 1935–1939 | S. H. Beaver | 1979–1982 | B. S. Morgan |
| 1945–1946 | N. V. Scarfe | 1983–1984 | K. Hilton |
| 1947–1950 | S. H. Beaver | 1985–1986 | D. R. Green |
| 1951–1955 | W. G. V. Balchin | 1987–1989 | N. Yates |
| 1956–1963 | R. C. Honeybone | 1990– | R. Chapman |
| 1964–1968 | P. Odell | | |

# Appendix I

## Honorary Librarians of the Association

| **Honorary Librarians** | | **Honorary Library and Information Officers** | |
|---|---|---|---|
| 1908–1914 | J. F. Unstead | 1977–1983 | T. W. Randle |
| 1918–1922 | H. O. Beckit | 1983–1985 | Elspeth Fyfe |
| 1930–1954 | T. C. Warrington | 1985–1991 | M. J. Shevill |
| 1955–1973 | L. J. Jay | 1991– | K. Luker |
| 1974–1977 | D. B. Grigg | | |

# Appendix J
## Honorary Members of the Association

Although introduced as early as 1901 when Dr. H. R. Mill was made an Honorary Member this method of acknowledging exceptional services to the Association appears to have lapsed until resuscitated during Professor Alice Garnett's Secretaryship. It is likely that the Association was unable to afford this particular luxury in its early years.

| | | | |
|---|---|---|---|
| 1901 | Dr. H. R. Mill | 1985 | Rosemary Robson |
| 1965 | Edgar Osbert Giffard | 1986 | Lord Nathan |
| 1968 | Charles B. Thurston | 1987 | John Old |
| 1978 | W. R. A. Ellis | 1988 | D. Riley |
| 1980 | Emeritus Professor W. G. V. Balchin | | A. J. Hunt |
| | Emeritus Professor Alice Garnett | 1990 | Dr. Gladys Hickman |
| | Alan D. Nicholls | | David Money |
| 1981 | Emeritus Professor S. H. Beaver | 1991 | Professor S. Gregory |
| | L. J. Jay | | Professor J. A. Patmore |
| | J. A. Morris | | G. M. Lewis |
| 1982 | Emeritus Professor R. W. Steel | 1992 | Professor Norman Graves |
| 1983 | R. S. Barker | | |
| | Emeritus Professor Norman Pye | | |
| | Emeritus Professor M. J. Wise | | |

# Appendix K

## Concise Biographies of Geographers contributing to the Association's History

**Stanley Beaver**, 1907–1984. Graduated UCL 1928 Honours Geography. Lecturer and Reader LSE, 1929–1951. Foundation Professor of Geography University College of North Staffordshire (subsequently Keele University) 1951 until retirement. Major contributor to Economic Geography of British Isles. Hon. Conference organiser GA, 1935–39 and 1947–50. President GA, 1966.

**Henry Oliver Beckit**, 1875–1931. Balliol College Oxford. Invited by Professor Herbertson to return to School of Geography, Oxford as an assistant 1908. Directed School after Herbertson's death 1915. Reader 1919. Acted as Hon. Editor of GA Journal, 1916–17 and Hon. Librarian, 1918–22.

**Rex A. Beddis**, 1932–1988. Monmouth School. Oxford Honours Geography. Taught in comprehensives at Putney and Holland Park, then Avery Hill College. Appointed co-director of Schools Council project 'Geography for the Young School Leaver', 1970. Subsequently Senior Adviser on Avon LEA. Author of numerous textbooks and had wide influence on teaching of school geography. Active in GA on Secondary Schools Committee and held post of Publications Officer, 1976–80.

**Emrys G. Bowen**, 1900–1984. Queen Elizabeth Grammar, Carmarthen and University College, Aberystwyth. Hon. Geography 1923. Asst. Lecturer Aberystwyth 1929 rising to Gregynog Professor, 1946 to 1968. Leading authority on Welsh Geography. GA Trustee, 1955–73, President, 1962.

**Leonard Brooks**, 1884–1964. Graduated Cambridge with Diploma in Geography 1910. Joined William Ellis School, North London, 1912. Pioneered concept of the school geography room (later laboratory). Prolific writer of school textbooks. Joined LCC Inspectorate 1920, became Divisional Inspector, 1929–1946. Founder-member of Ship Adoption Society. Served with RGS (Vice-President, Hon. Secretary, Hon. Treasurer, during 1939–1960). President GA 1951. Did a great deal to harmonise relations between GA and RGS.

**Robert Neal Rudmose Brown**, 1879–1957. Dulwich College. Aberdeen University trained as botanist and marine zoologist. Much influenced by Patrick Geddes. Travelled widely as explorer—Arctic, Antarctic, Burma. Lecturer in Geography, Sheffield University, 1908, Professor 1931–45. Pioneer advocate for physical basis and fieldwork in geography. President, Section E BA, IBG, Arctic and Antarctic Clubs. President, Sheffield Branch GA.

**Patrick Walter Bryan**, 1885–1968. Irish origins. Graduated in Geography at LSE after several years in business. Foundation Lecturer in Geography at University College of Leicester, 1922. Then for 22 years Vice-principal Head of Geography and Economics, and Dean of Faculty of Arts at Leicester. Skilled photographer and pioneer in use of visual aids for teaching. Wrote standard textbook on North America with Rodwell Jones. Professor at Leicester, 1953–54. President GA, 1957.

**C. C. Carter**. Died 1949. Master at Marlborough College. With support of Sir Cyril Norwood (Headmaster) made a name for himself with outstanding experimental lessons in Geography. Pioneer of new methods. Early textbook author. President GA, 1939–41.

**Vaughan Cornish**, 1862–1948. St. Paul's School. University of Manchester. D.Sc. in Chemistry. Science master then Director of Technical Education for Hampshire. Retired 1895 to devote himself to scientific travel and geographical research. Numerous books resulted on physical geography, scenery, strategic geography. President, Section E BA, 1923. Active supporter of GA and President 1928.

**Frank Debenham**, 1883–1965. Australian geologist attached to Scott's Antarctic expedition of 1910. Went to Cambridge to write up fieldwork. Served WW1, returned to Cambridge as Lecturer in Surveying and Fellow of Gonville and Caius. Reader in Geography 1928 and Professor 1931. Founder and Director of Scott Polar Research Institute, 1925–1946. Created post WW1 Cambridge Department of Geography. Innovative thinker. President GA, 1952.

**B. Bentham Dickinson**. Died 1941. Generally regarded as Founder of the Geographical Association as the organisation emerged from his efforts to arrange lantern slide lending. Master at Rugby School. Worked with A. W. Andrews in creating Diagram Slide Company. Association's first Hon. Secretary, 1893–1900. President GA, 1930.

**Kenneth C. Edwards CBE**, 1904–1982. University College Southampton, 1922–25. Lecturer, University College, Nottingham, 1926 and then Reader, Professor until retirement 1970. Active with Le Play Society. WW2 service with NID and Ministry of Town and Country Planning. Editor of GA series *British Landscapes through Maps*. Founded and edited *East Midland Geographer* 1954. Saw 50 years' service with Nottingham Branch of GA. President, Section E BA, 1959, IBG, 1960 and GA, 1963. CBE, 1970.

**James Fairgrieve**, 1870–1953. Glasgow High School to University College, Aberystwyth, then Jesus College, Oxford, 1891. Maths degree 1895. Teaching, eventually becoming Geography Master at William Ellis Grammar School, 1907. Joined GA and was destined to give 47 years of devoted service. From William Ellis School to London Day Training College as Lecturer in 1912, then Reader from 1931. Member of Committee and Council GA, 1909–1953, President GA, 1935. Pioneer advocate of land use study, local climatology, regional study, fieldwork, films, visual aids, political geography. A major 'backroom' contributor to the life of the Association.

**Herbert John Fleure, FRS**, 1877–1969. Guernsey origins. Aberystwyth graduate in Zoology, 1901. Became Assistant Lecturer, Aberystwyth in Zoology, Botany, Geology but had interests in Geography and Anthropology. Became Lecturer in Geography, 1908 and Professor of Zoology, 1910–1918, then Professor of Geography until move to Manchester as Professor of Geography, 1930–1944. President, Anthropological Section BA, 1926 and Geographical Section BA, 1932. Fellow of Royal Society, 1936. Hon. Secretary of GA, 1917–1946, President, 1948 and then Chairman of Council for 22 years.

**Douglas William Freshfield**, 1845–1934. Eton and University College, Oxford. Graduated History and Law. Private means enabled him to become a Victorian explorer, mountain climber and author—Alps, Himalayas, Caucasus in particular, thence to office in RGS. Developed an interest in geographical education and persuaded RGS to sponsor the John Scott Keltie investigation of 1884. Member of GA founding committee, 1893 and President GA, 1897–1911. Later President RGS, 1914–1917.

**Alice Garnett** 1903–1989. University College London, Honours Geography. Asst. Lecturer Sheffield 1924, stayed for 44 years becoming Professor, 1962 and Head of Dept., 1964. Joined GA, 1926 and served on Council, 1930–1940. Appointed Hon. Secretary GA, 1947 to 1967. Negotiated moves of GA from Manchester to Sheffield, Duke Street, 1950 and 343 Fulwood Road, 1964. Played significant role in development of GA. President IBG, 1966, President GA, 1968 and Chairman of Council, 1970–73.

**Sir Patrick Geddes**, 1854–1932. Scottish origins. Student of Biology. Taught Biology, Dundee University College, moved into Social Studies, Town and Regional Planning—planned some fifty cities in India and Palestine. Founded Outlook Tower in Edinburgh. Greatly influenced early Human Geographers. Escaped Presidency of GA as abroad so much but strong supporter. Son, Arthur Geddes (1895–1968) continued his father's work in the Dept. of Geography, Edinburgh.

**Andrew John Herbertson**, 1865–1915. Galashiels Academy. Edinburgh University. From a surveyor's office he went to geographical departments in the Universities of Freiburg-in-Breisgau, Montpelier and Paris: then Ben Nevis Observatory and editorial room of Buchan and Bartholomew's Atlas of Meteorology. Demonstrator in Botany at Dundee University College under Patrick Geddes, 1891–92: then from 1894–96, Owens College, Manchester and from 1899–1915 at School of Geography, Oxford. Head of School when Mackinder left in 1905 and personal chair followed, 1910. Hon. Secretary GA, 1900–1915 and first editor of *The Geographical Teacher*. Known for his concept of World Natural Regions.

**Sir William H. Himbury**, 1871–1955. Expert on cotton production. Founder-member of British Cotton Growing Association. Widely travelled in relation to cotton production and a pioneer of Commonwealth cotton. Governor for 35 years of Imperial College of Tropical Agriculture, Trinidad. Hon. Treasurer of GA, 1931–1955.

**Sir Thomas H. Holland, KCSI KCIE FRS**, 1868–1947. Canadian parentage. Royal College of Science, London. Geological Survey, India, 1890–1909. Professor of Geology, Manchester, 1909–1918. WW1 service in India. Rector, Imperial College, London, 1922–1929, then Vice-Chancellor, Edinburgh University until 1944. President of all scientific societies of which he was a member, including Geographical Association, 1938.

**J. L. Holland**. Died 1952. Elected to GA Committee in 1903 whilst master at St. Saviours and St. Olaves Grammar School. In 1905 became Secretary of Northamptonshire Education Committee and then Director of Education. Served GA until 1941. In 1929 promoted the publication of a Land Use map of Northamptonshire prepared by E. E. Field, from which sprang the First Land Utilisation Survey of Britain organised by L. Dudley Stamp.

**Osbert John Radcliffe Howarth OBE**, 1877–1954. Westminster and Christ Church, Oxford. History degree and Geography Diploma. 1904–1911, geographical assistant to Editor of *Encyclopaedia Britannica*. Then from 1909 to 1946 Secretary of British Association for the Advancement of Science. President, Section E BA, 1951. Wrote and edited numerous geographical books for OUP. Great advocate of Commonwealth. Life long interest in GA and became President in 1953.

**Geoffrey Edward Hutchings**, 1900–1964. Amateur field worker and naturalist who became an authority. Originally engineering drawing, Royal Dockyard, Chatham, then London with opportunity to attend Birkbeck College classes in Geology, Botany and Zoology. Teaching at Technical School, Baghdad, 1937, remaining abroad until 1945. Research officer Town and Country Planning, 1945–1947. Finally 1947 Warden of Juniper Hall for Council for Promotion of Field Studies. Created role of Warden at the Field Centres. Wrote several key field study manuals. President GA, 1961.

**John Riching James, OBE CB**, 1912–1980. Geography at King's College, London. Teaching, then Naval Intelligence, WW2. 1946, Research Officer, Ministry of Town and Country Planning, eventually rising to Chief Planner, 1961. In 1967, Professor of Town and Country Planning, University of Sheffield. Leading figure in planning world. Presidency of GA in 1969 quickly followed his return to academic work.

**Leslie J. Jay**. Died 1986. Geography at Birmingham University. Royal Navy in WW2. Taught at Wellingborough Grammar School before becoming Lecturer in Dept. of Education, Sheffield University. Hon. Librarian, GA, 1955–73 and wrote extensively on Library matters. Hon. Member GA, 1981.

**Llewellyn Rodwell Jones**, 1881–1947. Kingswood School, Bath, to London University. Initially science teacher then Lecturer in Railway Geography, 1913, Leeds University. After WW1 joined LSE and became Professor of Geography (after Sir Halford Mackinder) in 1925. Built up Joint School with King's College, London. Co-author with P. W. Bryan of classic *North America*. Provided much assistance for GA in relation to LSE.

**Sir John Scott Keltie**, 1840–1927. Scottish background. St. Andrews and Edinburgh Universities. Editorial posts with Chambers, 1861–71, Macmillan's, 1871–84, Statesman's Year Book, 1883–1927. Chosen by RGS as Inspector of Geographical Education and produced famous report of 1885. Secretary and Editor RGS, 1892–1915. President GA, 1914. Knighted 1918.

**David Leslie Linton**, 1906–1971. Joint School of Geography, King's College, London. Background of Physics, Chemistry, Geology, then Geography. Pre-WW2 Lecturer in Geography, Edinburgh University. Distinguished research in Geomorphology with S. W. Wooldridge. WW2 RAFVR in photographic intelligence. Chair of Geography, Sheffield University, 1945. Then Birmingham University, 1958. Honorary Editor GA, 1947–1964, President GA, 1964. Also President IBG, President, Section E BA, and FKC.

**Lionel William Lyde**, 1863–1947. Sedbergh School and Queen's College, Oxford. Classics graduate. School teaching, Edinburgh and Bolton School (Headmaster) until offered Professor of Economic Geography post at UCL in 1903 (at a salary of £50 per year). Remained until 1928. Prolific writer of school textbooks and said to have sold over 4 million. Profound influence on geography teaching in first decade of 20th century. President, North London Branch GA. Trustee, GA, 1914–1947.

**Colonel Sir Henry Lyons, FRS**, 1864–1944. Army background then Director, Geological Survey of Egypt, 1896–98, Director of Survey Dept., Egypt, 1898–1909, Head of Geography Dept., University of Glasgow, 1909–14, director of Meteorological Office, 1914–20, Director of Science Museum, 1920. Knighted 1926. Hon. Treasurer GA, 1926–28, President GA, 1929.

**Sir Halford Mackinder, PC**, 1861–1947. Gainsborough Grammar School, Epsom College, Christ Church, Oxford, read Natural Science, History, Law. University Extension lecturing on The New Geography led to RGS sponsored Readership in Geography at Oxford. Founded School of Geography, 1887. Remained at Oxford until 1905 although also Principal at Reading University College, 1892–1903 and Director, London School of Economics, 1903–08. Reader Geography, LSE, 1903–23 and Professor to 1925. From 1910 to 1922, Member of Parliament. Founder-member of GA, 1893, Chairman of Committee, then Council, 1908–47, Trustee, 1914–47, President, 1916. Profound influence on teaching of geography, wrote key texts at all levels.

**Ernest Cecil Marchant, CIE**, 1902–1979. St. John's College, Cambridge. Taught Oakham School. Accompanied J. A. Steers on Great Barrier Reef expedition and then remained in Australia teaching at Geelong Grammar. Joined Marlborough College, 1931, then Daly College, India, 1939. After WW2 returned UK and appointed HMI. Staff Inspector for Geography 1955. Considerable influence on teaching of Geography. Long connections with GA and President, 1967.

**John Story Masterman**, 1849–1931. Rugby School. Corpus Christi College, Oxford. Classics. Assistant Master, University College School, London, 1880–1894. Amateur naturalist and mountaineer led to interest in Geography and brought into GA by B. B. Dickinson, 1894. Acted as Assistant Hon. Secretary, 1894–1900 and Hon. Treasurer, 1900–1907.

**Hugh Robert Mill**, 1861–1950. University of Edinburgh, studied Chemistry but came under influence of James Geikie, Patrick Geddes, Sir John Murray and J. Y. Buchanan. Gave University Extension lectures on Physical Geography. Moved to London as Librarian, RGS, 1892 but resigned 1900 to take over British Rainfall Organisation. Became his life's work but he considered himself a Geographer. President, Section E BA, 1901, President Royal Meteorological Society, 1907. Wrote extensively on geographical topics, especially Polar Regions. Long connection with GA and President 1932.

**A. Austin Miller**, 1900–1968. University College, London, Geology, 1922. Moved into Geography and appointed Lecturer, Reading University, 1926, Professor 1943. International reputation for pioneer work in climatology and geomorphology. President IBG 1946–48, President, Section E BA 1956, President GA 1960.

**Joseph Acton Morris**. Died 1987. Taught Latymer School, Edmonton, 1926–62. Head of Geography, 1931 and Deputy Headmaster, 1952. Author of numerous textbooks on Geography and History. Long service on GA committees and President GA, 1965. Honorary Member GA, 1981.

**Sir John Linton Myres**, 1870–1954. From 1910–1939 Professor of Ancient History at Oxford but strong advocate of Geography. Intelligence Officer WW1. Greatly assisted GA in negotiations with Board of Education. Trustee GA, 1927–1954. President GA, 1925.

**Alan D. Nicholls**, 1910–1981. Truro School, Exeter and London Universities. LCC 1933, seconded Ministry of Supply, 1941–43. St. Clement Danes, Hammersmith, 1944 eventually following Leonard Suggate as Head of Geography. Keen field geographer. Long period of association with GA on Secondary Schools Committee and Chairman of Central London and Westminster Branch. Hon. Asst. Treasurer GA prior to becoming President GA, 1972.

**Percy Maude Roxby**, 1880–1947. Bromsgrove School. Christ Church, Oxford. Graduated 1903 in History. Then Diploma courses in Geography prior to Asst. Lectureship at Liverpool 1904. Lecturer in Geography, 1906 and Professor, 1917–1944. Became international expert on China. NID Intelligence in WW2. Associate Editor of Journal *Geographical Teacher*, 1916–32 and President GA, 1933.

**Sir E. John Russell, FRS**, 1872–1965. Aberystwyth, Science and Manchester, Chemistry, then Wye College as soil chemist. Succeeded Sir Daniel Hall at Rothamsted in 1912. Gifted research worker and environmentalist, world authority on pedology. Keen supporter of GA and Le Play Society—joint interest in Regional Surveys. Keen field worker and led many study tours abroad (up to age 86!). President Le Play Society, 1938–60, President British Association, 1949, President GA, 1923.

**Neville Scarfe**, 1908–1985. University College, London. Day Training College under Fairgrieve. Teaching then Lecturer, University College, Nottingham. In 1935 replaced Fairgrieve at London Institute of Education. Active member of Le Play Society. Strong advocate of visual aids and fieldwork. Hon. Conference Organiser GA, 1945–1946. Dean of Education at University Manitoba, 1951 and University of British Columbia, 1956.

**Sir Laurence Dudley Stamp, CBE**, 1898–1966. King's College, London—four years spectacular, beginning at age 15, taking three Intermediates and two Honours courses in Botany and Geology. WW1 service and then Lecturer in Geology KCL. Took first Honours BA in Geography in London University in 1921. Professor of Geography and Geology, Rangoon University, 1923. Returned to LSE in 1926 as Sir Ernest Cassel Reader in Economic Geography. Professor of Geography, 1945. Director of First Land Utilisation Survey of Britain, 1930s. Prolific writer of textbooks and dominated school market at home and overseas during inter-war period. Government adviser and International geographer. President IGU, 1952–56, President RGS, 1964–66, President, Section E BA, 1949, President IBG, 1956, President GA, 1950 and also Hon. Treasurer GA, 1955–66. CBE, 1946. Knighthood, 1965.

**James Alfred Steers, CBE**, 1899–1987. St. Catharine's College, Cambridge. Demonstrator in Geography, Cambridge, 1922, then Lecturer, 1927 and Professor, 1949–66. Leading coastal physical geographer and major figure in development of university geography. Key contributor to setting up of Nature Conservancy and National Parks, 1940s. Coastal adviser to National Trust, Government and Council for Europe. President GA, 1959. Strong supporter Public and Preparatory School group.

**Leonard S. Suggate**, 1889–1970. King's College and Day Training College, London. Later Geography master at St. Clement Danes School, 1917–49. Advocate of visual aids, geography rooms and map work. Chief examiner, London General and London Higher School certificates for many years. Initiated 'This Changing World' in *Geography*. Major influence in evolution of the subject—member of GA for nearly 60 years. President GA, 1955.

**Harry Thorpe, OBE**, 1913–1977. Chesterfield Grammar School, Durham University. Taught Accrington Grammar School, 1936–39. WW2 Major in Royal Engineers. Lecturer in Geography, University of Birmingham, 1946 rising to Professor and then Head of Department, 1971. Holistic historical geographer. OBE 1976 for work on the allotment movement for Ministry of Land and Natural Resources. Chairman of Birmingham Branch GA (one time largest in country) from 1953. Also President of Worcester Branch and involved with Coventry and Wolverhampton Branches. President of GA, 1974.

**Charles B. Thurston**, Died 1969. Graduated in Natural Sciences. Trained at Borough Road College. Taught Geography, Kilburn Grammar School. WW1 served as Meteorologist. Returned to Kilburn Grammar then Headmaster, Isleworth County School. Widespread influence through his textbooks. Chairman, Secondary Schools Committee of GA for 25 years, 1927–1952. Invited to be President for 1955 but had to decline owing to his wife's illness. Made Honorary Member GA, 1968.

**John Frederick Unstead**, 1876–1965. Pupil teacher London then Day Training College, Cambridge, 1895. Came under influence of Mackinder and Herbertson and in 1905 appointed Lecturer in Geography at Goldsmith's College. Then Birkbeck College where made Professor 1921 until 1930 when he retired to concentrate on writing. Author of numerous textbooks and particularly interested in regional geography. Honorary Librarian and 'Correspondence Secretary' GA, 1908–1914.

**Thomas Cotterill Warrington**, 1869–1963. Newcastle under Lyme Grammar School. University College of Wales, Aberystwyth, reading Chemistry, Maths, Physics, Proceeds to Jesus College, Oxford, graduating in Natural Science, 1894. Science teaching Caernarvon, then Headmaster Leek Grammar School, 1900. Elected to GA Council, 1930 and served in many aspects continuously thereafter. Regular contributor to *Geography*, OS Maps Officer for GA, Honorary Librarian GA, 1930–1954. President GA, 1942–45.

**Sidney William Wooldridge, CBE, FRS**, 1900–1963. King's College, London, graduate in Geology. Asst. in Dept. of Geology, King's 1922, then Lecturer in Geography Joint Dept., 1927. Professor of Geography, Birkbeck College, 1944, then King's College, 1947. Major advocate of Physical Geography as basis of geographical study. Founding figure in British Geomorphology. Profound influence through Field Studies Council on development of field study in Geography. President IBG, 1949–50, President, Section E BA, 1950, President GA, 1954. CBE, 1954, for work on Sand and Gravel Advisory Committee.

# Appendix L

## Chairs of Geography in British Universities

One indicator of the growth and consolidation of geography in Great Britain in the twentieth century is the appearance of chairs in the subject in British Universities. At the beginning of the century no chairs existed at all. As a result of the combined efforts of the Royal Geographical Society and the Geographical Association however chairs gradually came to be established, at first very slowly as much depended on endowments (e.g. Gregynog at Aberystwyth, John Rankin at Liverpool) or other financial support. It was not until the 1950s that nearly all the then universities could claim a chair in Geography. Thereafter growth was more rapid: the 1960s were marked by the introduction of second chairs, and the 1970s by third and even fourth chairs in the larger departments. Expansion came to a halt in the early 1980s however with the Government's financial squeeze, but might well resume in the 1990s with the further change in policy—both university and the now upgraded earlier polytechnic institutions are being encouraged to expand numbers once again.

The following list covers the period 1900 to 1980:

|  | |
|------|---|
| 1903 | UCL (Lyde) |
| 1907 | Reading (Dickson) |
| 1910 | Oxford (Herbertson) |
| 1917 | Aberystwyth (Fleure), Liverpool (Roxby) |
| 1922 | Birkbeck (Unstead), LSE (Mackinder) |
| 1925 | LSE (Rodwell Jones) |
| 1926 | Southampton (Rishbeth) |
| 1927 | Exeter (Lewis) |
| 1928 | UCL (Fawcett) |
| 1930 | Aberystwyth (Forde), Birkbeck (Eva Taylor), Manchester (Fleure) |
| 1931 | Cambridge (Debenham), Edinburgh (Ogilvie), Sheffield (Rudmose Brown) |
| 1932 | Oxford (Mason) |
| 1933 | Bristol (Jervis) |
| 1943 | Newcastle (Daysh), Reading (A. Miller) |
| 1944 | Birkbeck (Wooldridge), Leeds (Williamson), Manchester (Fitzgerald) |
| 1945 | Belfast (Estyn Evans), Liverpool (Darby), LSE (Stamp), Sheffield (Linton) |
| 1946 | Aberystwyth (Bowen) |
| 1947 | Birkbeck (East), Glasgow (Stevens), KCL (Wooldridge) |
| 1948 | Bedford (Manley), Birmingham (Kinvig), Exeter (Davies) |
| 1949 | Cambridge (Steers), LSE (Buchanan), Nottingham (Edwards), UCL (Darby) |
| 1950 | Keele (Beaver), Liverpool (Smith) |
| 1951 | Aberdeen (O'Dell), Leeds (Peel) |
| 1953 | Glasgow (R. Miller), Leicester (Bryan), Manchester (Crowe), Oxford (Gilbert) |
| 1954 | Edinburgh (Wreford Watson), Hull (King), Leicester (Pye), Southampton (Monkhouse), Swansea (Balchin) |
| 1955 | QMC (Smailes) |
| 1956 | Durham (W. B. Fisher) |
| 1957 | Bristol (Peel), Liverpool (Steel) |
| 1958 | Birmingham (Linton), Hull (Wilkinson), Leeds (Dickinson), LSE (Wise) |
| 1959 | Sheffield (C. A. Fisher) |
| 1961 | LSE (Emrys Jones) |
| 1962 | Sheffield (Alice Garnett), UCL (Mead) |
| 1963 | Durham (Bowen Jones), KCL (Hare), SOAS (C. A. Fisher), Sussex (Elkins) |
| 1964 | Aberystwyth (Kidson), Bedford (Monica Cole), Lancaster (Manley), LSE (Harrison-Church), Newcastle (House), Sheffield (Waters) |
| 1965 | Aberdeen (Walton), Birkbeck (Henderson), Edinburgh (Coppock), Newcastle (Conzen), Reading (T. G. Miller) |
| 1966 | Bristol (Haggett), Cambridge (Darby), KCL (Pugh), Swansea (Oliver) UCL (Brown), UCL (Wheatley) |
| 1967 | Aberdeen (Mellor), Dundee (S. J. Jones), Leeds (Birch), Nottingham (Osborne), Southampton (Bird), Strathclyde (Howe), Ulster (Oldfield) |
| 1968 | Aberystwyth (Carter), Durham (Clarke), Oxford (Gottman), Reading (Hall), Sheffield (S. Gregory) |
| 1969 | Aberdeen (Ritchie), Belfast (Kirk), Exeter (Ravenhill), Nottingham (Cuchlaine King), Reading (Savigear) |
| 1970 | Birkbeck (Eila Campbell), Keele (Rodgers), Lampeter (D. Thomas) Open (Learmouth), Leeds (Wilson), SOAS (Hodder), Swansea (Greenwood) Liverpool (Lawton) |
| 1971 | Exeter (Straw), KCL (W. B. Morgan), Manchester (Rodgers), QMC (Rawstron) |

| | |
|---|---|
| 1972 | Bristol (Chisholm), UCL (Lowenthal) |
| 1973 | Birmingham (Temple), Birmingham (Moss), Hull (Patmore), Leeds (Kirkby), Manchester (Chandler), QMC (D. M. Smith), Salford (White), Ulster (Barbour) |
| 1974 | Aberdeen (Stephens), Cambridge (Chorley), Keele (Dwyer), Lancaster (J. J. Johnson), Leeds (Glanville Jones), Manchester (Freeman), St. Andrews (Proudfoot), Sheffield (R. J. Johnson) |
| 1975 | Bedford (Cooke), Dundee (Caird), Leicester (Paterson), Newcastle (Simpson), Newcastle (Goddard), Nottingham (Cole), Liverpool (Oldfield) |
| 1976 | Cambridge (Chisholm), Glasgow (Thompson), Glasgow (Tivy), Southampton (K. J. Gregory) |
| 1977 | Glasgow (Petrie), Liverpool (Prothero), Manchester (Robson), Ulster (Langlands) |
| 1978 | Birmingham (D. Thomas), Bristol (Simmons), Hull (Appleton), IEL (Graves), Lampeter (Beaumont), Swansea (Stephens) |
| 1979 | Aberdeen (Ritchie), Loughborough (Butlin), Manchester (Douglas), Salford (Moss) |
| 1980 | UCL (Manners) |

Abbreviations:
Bedford—Bedford College, University of London
Birkbeck—Birkbeck College, University of London
IEL—Institute of Education, University of London
KCL—King's College, University of London
LSE—London School of Economics, University of London
QMC—Queen Mary College, University of London
UCL—University College, University of London

# Appendix M

## Publications of the Association

The main publications of the Association are its three journals, each of which now appears quarterly.

*Geography*, which began as *The Geographical Teacher* in 1901, had its name change in 1927, and has appeared without a break throughout the present century despite the two World Wars.

*Teaching Geography* was first published in 1975 in response to requests from the profession for a journal more specifically related to the practical problems of those at 'the chalk face'.

*Primary Geographer* followed in 1989 in order to meet the needs of those involved in teaching geography at the primary level.

The Honorary Editors of the journals are listed in Appendix E and details of the foundation, content and evolution of the journals will be found in the main text.

———————

The Association has also published numerous occasional papers, pamphlets, booklets, handbooks, books, library catalogues, landform guides and fieldwork guides to assist the geography teacher. No royalties or fees have accrued to authors or editors of these publications and they have without exception given their time, energy and expertise freely for the benefit of their colleagues.

The following is a complete list of all the Association's publications other than reprints from journals.

# General Publications

## 1910s

Guide to Geographical Books and Appliances. H. R. Mill. 1910. (First edition, RGS 1897, Hints to teachers and students on the choice of geographical books for reference and reading.)

The Trilogy of the Humanities in Education. An address given to the Tredegar and District Co-operative Society. H. J. Fleure. 1918.

A scheme for a First Course in Geography. (With an additional syllabus for standards I and II.) 1919.

Short List of Books, Atlases and Apparatus Useful in the Teaching of Geography. 1919.

Geography in Education, by members of the Geographical Association. Report of the results of discussions held during the short course in geography for secondary teachers, organised for the Board of Education at the University College of Wales, Aberystwyth. 1919.

## 1920s

Suggestions for Local Studies of British Antiquity. H. J. Fleure. 1920.

The Far Eastern Question in its Geographical Setting. Percy M. Roxby. 1920.

Historical Geography of the Wealden Iron Industry. Miss M. C. Delany. 1921.

Historical Geography of Ireland. W. Fitzgerald. 1926.

Agricultural Geography of the Deccan Plateau. Miss E. Simkins. 1927.

## 1930s

Suggestions for the Teaching of Geography, 1930.

Handbook for Geography Teachers. Miss D. M. Forsaith. 1932.

Suggestions for the Teaching of Local Geography. 1934.

The Turnpike Roads of Nottinghamshire. Arthur Cossons, 1934. Joint publication, the Geographical Association and the Historical Association. To be extended and reprinted by Nottinghamshire Archives Office, 1993.

Local Studies. Prepared under the auspices of members of the Geographical Association's Standing Committee for Geography in Secondary Schools. First edition. 1939. (Revised editions, 1946 and 1949.)

## 1940s

Geography in the Primary School. A report prepared by a special sub-committee of the Primary School Group and the Training College Group of the Geographical Association. 1949. (Revised edition, March 1953 and 1955.)

## 1950s

The making of Geography Teaching Films. Prepared by the Films Committee of the Geographical Association. (Published by the National Committee for Visual Aids in Education.)

Geography in the Secondary School. With special reference to the secondary modern school. A report prepared at the request of the Executive Committee of the Geographical Association. E. W. H. Briault and D. W. Shave. 1952. (Second edition 1955, revised and reprinted 1960, 1962, 1964 and 1966.)

The Geography Room in a Secondary School. P. R. Heaton. 1954.

Teaching Geography in Junior Schools. 1959. Revised 1962, 1965.

## 1960s

Exercises on Ordnance Survey Maps. 1. One-inch Tourist Map, The Lake District. Edited by A. D. Nicholls. 1960.

Exercises on Ordnance Survey Maps. 2. One-inch Sheet 90, Yorkshire Dales, Wensleydale. Edited by A. D. Nicholls. 1960.

Sample studies. 1962. Revised 1966.

Local Study in the CSE Examination. R. A. Beddis. 1965.

The Geography Room and its Equipment. R. Cole. 1968.

Asian sample studies. 1968.

## 1970s

Geography in primary schools. Edited by Miss P. H. Pemberton, 1970.

Geography in secondary education. N. J. Graves. 1971.

Geography in Education. A Bibliography of British Sources. 1870–1970. Clare T. Lukehurst and N. J. Graves. 1972.

Degrees in Geography in Britain, a Guide for the Intending Student. P. R. Mounfield and David Thomas. 1974. (Published with the Institute of British Geographers.)

The Sir Dudley Stamp Memorial Index to Geography, Volumes 1 to 54, 1901–1969. L. J. Jay and Hilary Todd. 1974.

## 1980s

Geographical Education in Secondary Schools. Norman J. Graves. 1980.

Computer Assisted Learning in Geography. Ifan D. H. Shepherd, Zena A. Cooper and David R. F. Walker. 1980.

Geography into the 1980s. Proceedings of a conference held in Oxford in March 1980 on the contribution of three Schools Council Projects to geography in secondary education. Edited by Eleanor Rawling. 1980.

Geographical Work in Primary and Middle Schools. Edited by David Mills. 1981. (Revised edition, edited by David Mills and Wendy Morgan, 1988.)

Language, Teaching and Learning. 2: Geography. A report of a Working Group set up by the Education Standing Committee of the Geographical Association. Edited by Michael Williams. 1981. (Published by Ward Lock Educational.)

GYSL with the Disadvantaged. Edited by David Boardman. 1981.

Geography with Slow Learners. Edited by David Boardman. 1982.

The Good, the Bad and the Ugly. Joy A. Palmer and Margaret J. Wise. 1982.

New Leads in Geographical Education. Edited by Keith Orrell and Patrick Wiegand. 1982.

Register of Research in Geographical Education. Graham Corney. 1982.

The Contribution of Geography to 17+ Courses. Report by the Geographical Association's Working Party on Examinations. 1982.

Teaching Geography to Less Able 11–14 Year-olds. Report by the Working Group on New Techniques and Methods in Teaching Geography. 1982.

Patterns on the Map. 1982.
No. 1    Introductory Handbook. Alice Coleman and Simon Catling.
No. 2    Plymouth and Merthyr Tydfil. Alice Coleman, H. A. Sandford, John Bale and Graham Humphrys.
No. 3    Leeds and Rosedale. Alice Coleman, John Bale and Michael Hewitt.
No. 4    Sevenoaks and Gravesend. Alice Coleman, Inga Feaver and Raymond Pask.

England and Wales '81. A. G. Champion. 1983.

The Third World: Issues and Approaches. Edited by John Bale. 1983.

Evaluation and Assessment in Geography. Edited by Keith Orrell and Patrick Wiegand. 1983.

Geography Graduate Careers. Edited by R. Hebden. 1983.

Geography and Careers: The School Leaver. T. W. Randle. 1983.

Designing and Teaching Integrated Courses. Michael Williams. 1984.

Geography, Schools and Industry. Edited by Graham Corney. 1984.

Teaching Slow Learners Through Geography. Edited by Graham Corney and Eleanor Rawling. 1984.

Geography Software: Geographical Association/ MEP Project. Project Director, David Walker. Author, Howard Midgley. 1984.

Worldwise Quiz Book, No. 1. 1984.

List of Geography Microcomputer Software. Peter Fox and members of the Educational Computing Working Group. 1984.

Geographical Futures. Russell King. 1985.

Geographical Education for a Multi-cultural Society. Rex Walford. 1985.

Worldwise Quiz Book, No. 2. 1985.

Perspectives on a Changing Geography. Edited by Ashley Kent. 1985.

Local Studies 5–13. Suggestions for the Non-specialist Teacher. 1985. (Updated 1991.)

Geography Beyond A-level. Edited by G. Malcolm Lewis. 1985.

Handbook for Geography Teachers. Edited by David Boardman. 1986.

The Role and Value of New Technology in Geography. Edited by Peter Fox and Andrea Tapsfield. 1986.

Profiling in Geography. Edited by Norman J. Graves and Michael Naish. 1986.

Geography and Careers: An Information Pack for Teachers. Edited by Paul Birchall. 1986. (Revised 1991.)

Computers in Action in the Geography Classroom. Edited by Ashley Kent. 1987.

Starting to Teach Geography. L. J. Jay. 1987.

Worldwise Quiz Book, No. 3. 1987.

Geography and Pre-Vocational Education. Edited by John Bale. 1988.
No. 1    Geography and TVEI
No. 2    Teachers Talking about Geography, TVEI and Place Studies
No. 3    Geography and CPVE
No. 4    GYSL-TRIST: An example of Geography in pre-vocational education.

A Case for Geography. Edited by Patrick Bailey and Tony Binns. 1988.

Low Attainers and the Teaching of Geography. Jacqueline L. Dilkes and Aubrey Nicholls. 1988. (Jointly with the National Association for Remedial Education.)

Worldwise Quiz Book, No. 4. 1988.

Managing the Geography Department. Edited by Patrick Wiegand. 1989.

Methods of Presenting Fieldwork Data. P. R. St. John and D. A. Richardson. 1989.

Geography in the National Curriculum: A Viewpoint from the Geographical Association. Edited by Richard Daugherty. 1989.

Geography through Topics in Primary and Middle Schools (including the application of information technology). 1989. (Joint publication with the National Council for Educational Technology.)

Some Issues for Geography within the National Curriculum. Edited by Simon Catling. 1989.

## 1990s

Selling Geography. Edited by Ashley Kent. 1990.

INSET Pack. Edited by James Hindson and Jacqueline Dilkes. 1990.

Images of Poland—Urban and Rural Studies. Edited by Patrick Bailey and Peter Fox. 1990.

Primary Geography Matters. Conference Proceedings, 18 April 1990.

Methods of Statistical Analysis of Fieldwork Data. D. A. Richardson and P. R. St. John. 1990.

Geography Outside the Classroom. 1990.

Issues in Mapping the Future of School Geography 5–16. Edited by by Simon Catling. 1990.

Teaching Economic Understanding Through Geography. Edited by Graham Corney. 1991.

Worldwise Quiz Book, No. 5. 1991.

The Word Puzzle Book. Andrew Dalwood. 1991.

Geography, IT and the National Curriculum: Teacher's and Case Studies Booklets. 1991. (Joint publication with the National Council for Information Technology.)

Planning for Key Stage 2. Wendy Morgan. 1991.

Answers . . . To Teachers' Questions. A Guide to the Geographical Association's 'Geographical Work in Primary and Middle Schools'. Wendy Morgan. 1991.

Geography and History Through Stories: Key Stage One. A. Gadsden. 1991. (Joint publication with the Cheshire Primary Humanities Team.)

Planning for Geography for Pupils with Learning Difficulties. Judy Sebba. 1991.

Planning for Key Stage 1. Jill Raikes. 1991.

Primary Geography Matters—Inequalities. Conference Proceedings. 1991.

Geography and History in the National Curriculum. Heather Norris Nicholson. 1992.

Talking about Geography: The Work of Geography Teachers in the National Oracy Project. Edited by Roger Carter. 1992.

Teaching About Places: East Kent Examples. Harry Mountford. 1992.

Geography National Curriculum—Programmes of Study: Try this Approach. E. Rawling. 1992.

Placing Places: 201 Stimulating Ways to Introduce Locational Knowledge to Children. Simon Catling. 1992.

Development Education in the Geography National Curriculum (Introduction and Resources). Marjorie Drake. 1992.

Water in the Environment. Rachel Bowles. 1992.

Earth in the Environment. Rachel Bowles. 1992.

Soil, Plants and the Environment. Rachel Bowles. 1992.

Teaching Economic Understanding Through Geography. Edited by Graham Corney. 1992.

A Student's Guide to Studying Geography in the Sixth Form. John Chubb. 1992.

Primary Schools, Geography and the National Curriculum. Michael Naish. 1992.

Earth Science Education Forum Directory. 1992.

(Joint publication with the Geological Society.)

Where is the School Exactly? A Practical Guide for Those Returning to Supply or Full-Time Teaching in Primary Schools. Anne Cox. 1992.

Place: A Practical Guide to Teaching About Places. A Collaborative Project between Cheshire, Cumbria, Lancashire, Shropshire, Staffordshire LEAs and the Universities of Keele and Liverpool. 1992.

Primary Geography Matters—Change in the Primary Curriculum. Conference proceedings. 1992.

St. Lucia Photopack. 1992. (A joint publication with CWDE.)

Ladakh Photopack. 1992.

Worldwise Quiz Book, No. 6. 1992.

The Worldwise Quiz—Computer Quiz. 1992.

**Theme Issues reprinted from Geography, This Changing World.**

Industrial UK Up-to-Date. 1984.

Socialist Development in the Third World. 1987.

Geographical Perspectives on the Crisis in Africa. 1988.

Forty Years of the People's Republic of China. 1989.

The Changing Face of Eastern Europe and the Soviet Union. 1990.

Industrial UK Up-to-Date. 1990.

The Human Impact on the Environment. 1991.

British Passenger Transport into the 1990s. 1992.

**Library Catalogues**

Catalogue of reference library. March 1916.

Catalogue of the postal reference library of the Geographical Association. 1923.

Catalogue of the postal reference library of the Geographical Association. 1924.

Subject index and author catalogue of the postal reference library of the Geographical Association. 1925.

Subject index and author catalogue of the additions to the postal reference library of the Geographical Association made between May 1925 and May 1926. 1926.

Subject and author catalogue of the reference library of the Geographical Association. 1933.

Asia. L. J. Jay. 1955.

Africa. L. J. Jay. 1957.

North and Latin America. L. J. Jay. 1958.

Australia, New Zealand and the Oceans. L. J. Jay. 1961.

Library Catalogue: The Americas. L. J. Jay. 1968.

**British Landscapes Through Maps** (K. C. Edwards, editor)—Commenced 1960

No. 1   The English Lake District. F. J. Monkhouse. 1960. (2nd revised edition, 1972)

No. 2   The Yorkshire Dales. C. A. M. King. 1960.

No. 3   Guernsey. H. J. Fleure. 1961.

No. 4   The Chilterns. J. T. Coppock. 1962.

No. 5   Snowdonia. C. Embleton. 1962.

No. 6   Merseyside. R. Kay Gresswell and Richard Lawton. 1964.

No. 7   The Scarborough District. Cuchlaine A. M. King. 1965.

No. 8   The Doncaster Area. Bryan E. Coates and G. Malcolm Lewis. 1966.

No. 9   Cornwall. W. G. V. Balchin. 1967.

No. 10  East Kent. Alice M. Coleman and Clare T. Lukehurst. 1967.

No. 11  The Oxford and Newbury Area. P. D. Wood. 1968.

No. 12  Dartmoor. D. Brunsden. 1968.

No. 13  The Strathpeffer and Inverness Area. Alan Small and John S. Smith. 1971.

No. 14  The Norwich Area. Patrick Bailey. 1971.

No. 15  The Loch Linnhe District. A. J. Cruickshank and A. J. Jowett. 1972.

No. 16  The Fishguard and Pembroke Area. B. S. John. 1972.

No. 17  The Worcester District. Brian H. Adlam. 1974.

No. 18  Cheltenham and Cirencester. Michael Naish. 1978.

No. 19  The Potteries. S. H. Beaver and B. J. Turton. 1978.

**Teaching Geography Occasional Papers** (edited by N. J. Graves)—Commenced 1967

No. 1   A Topical List of Vertical Photographs in the National Air-Photo Libraries. Alan D. Walton. 1967.

No. 2   Do-it-yourself Weather Instruments. P. D. Hookey. 1967.

No. 3   The Uses of a Revolving Blackboard in Geography Teaching. J. A. Bond.
A Technique of Using Screen and Blackboard to Extract Information from a Photograph. R. A. Beddis.
Producing a Slide Set with Commentary from Elementary Fieldwork. E. F. Thornton. 1968.

No. 4   Geography Books for Sixth Forms. L. J. Jay. 1968.

No. 5   Teaching Aids on Australia and New Zealand. D. G. Mills. 1969.

No. 6   Fieldwork Using Questionnaires and Population Data. R. J. P. Newman. 1969.

No. 7   Hardware Models in Geography Teaching. E. W. Anderson. 1969.

No. 8   Exercises on Ordnance Survey Maps and Extracts: Stow on the Wold. Michael Naish. 1969.

No. 9   Exercises on Ordnance Survey Maps and Extracts: Bishop Auckland. Rex Walford. 1969.

No. 10  Exercises on Ordnance Survey Maps and Extracts: Calne; Stirling. Molly Long and B. S. Roberson. 1969.

No. 11  Hypothesis Testing in Field Studies. D. P. Chapallaz, P. F. Davis, B. P. Fitzgerald, N. Grenyer, J. Rolfe and D. R. F. Walker. 1970.

No. 12  A Market Survey: Techniques and Potentialities. M. M. Baraniecki and D. M. Ellis. 1970.

No. 13 Latin America's Economic Situation: The Use of Rank Correlation Coefficient. R. J. Robinson. 1970.

No. 14 Traffic Study as Quantitative Fieldwork. Alastair Morrison. 1970.

No. 15 An Introduction to the Analysis of Road Networks. W. V. Tidswell. 1971.

No. 16 Geography Books for Sixth Forms. L. J. Jay. 1971.

No. 17 An Instructional Exercise in Industrial Location: The Electricity Industry. G. B. Hardy and R. L. White. 1972.

No. 18 Some Aspects of the Study and Teaching of Geography in Britain: A review of recent British research. M. C. Naish. 1972.

No. 19 A Visit to Kew Gardens as Biogeographical Teaching. E. M. Yates and R. Robson. 1973.

No. 20 An Approach to Fieldwork in Geomorphology: The Example of North Norfolk. W. A. Kent and K. R. Moore. 1974.

No. 21 Drainage Basin Instrumentation in Fieldwork. E. W. Anderson. 1974.

No. 22 Teaching the Commercial Geography of a Port. B. M. Palmer. Three Dimensional Models: Their Construction and Use. M. C. D. Rudd. 1974.

The following were published under the new editor: M. C. Naish

No. 23 Practical Biography. Rona Mottershead. 1974.

No. 24 A Further Exercise in Industrial Location: The Influence of Behavioural Factors Upon the Distribution of Industry—The Broiler Industry. G. B. Hardy and R. L. White. 1974.

No. 25 Hydrology for Schools. Darrell Weyman and Charles Wilson in co-operation with Pat Cleverley and David Ingle Smith. 1975.

No. 26 Motorway. E. Rawling. 1976.

No. 27 Analysis of Land Use Data: A Practical Exercise. R. Daugherty. 1976.

No. 28 Geography and Football: The Use of Ideas from Football in the Teaching of Geography. J. Bale and D. Gowing. 1976.

No. 29 Landscape Drawing: A Neglected Aspect of Graphicacy. R. L. Simmons and J. K. Mears. 1977.

No. 30 Some Aspects of the Study and Teaching of Geography in the United States: A Review of Current Research, 1965–1975. R. N. Saveland and C. W. Pannell. 1978.

No. 31 Geography Through Topics in Primary and Middle Schools. J. R. Cracknell. 1979.

No. 32 Mental Maps: Resources for Teaching and Learning. Joseph P. Stoltman. 1980.

No. 33 The Geography of Outdoor Recreation: Suggestions for Practical Exercises and Fieldwork. C. J. Bull and P. A. Daniel. 1982.

**Bibliographic Notes** (edited by T. W. Randle)
Commenced 1979

No. 1 Assessment and the Geography Teacher. Melvyn Jones. March, 1979.

No. 2 Mixed Ability Group Work in Secondary School Geography. GA Working Group on New Techniques and Methods with the assistance of R. Daugherty and R. Johnson. March, 1979.

No. 3 Population Geography at A-level. Melvyn Jones and Bill Hornby. November 1979.

No. 4 The European Economic Community. Andrew Convey. February 1980.

No. 5 Geography and Integrated Studies in Secondary Schools. Richard Daugherty. March 1980.

No. 6 The Study of Recreation and Leisure in School and College. Environmental Education Working Group of the GA. March 1980.

No. 7 Fieldwork Techniques in Schools. J. Alan Taylor. March 1980.

No. 8 Agricultural Geography. Vincent Tidswell. January 1981.

No. 9 Mapwork in Primary and Middle Schools. Simon Catling. March 1981.

No. 10 The Geography of Development for 16–19 Year-olds. Elspeth Fyfe and Bill Hornby. April 1981.

No. 11 Games and Simulations in Geography Teaching. Rex Walford. June 1981.

No. 12 The United States of America. J. Arwel Edwards, Terry Morgan and Clive Edwards. June 1981.

No. 13 Industrial (Manufacturing) Geography. Colin Read. July 1981.

No. 14 Cognitive Studies in Geography. Miriam J. Boyle. December 1981.

No. 15 The Two Germanies. Verna Freeman. March 1982.

No. 16 France. Christopher Rogers. April 1982.

No. 17 The Soviet Union and Eastern Europe. Verna Freeman and Ralph Hebden. April 1982.

No. 18 Norden: The Scandinavian States. Professor W. R. Mead. April 1982.

No. 19 The European Economic Community (2nd edition) Andrew Convey. April 1982.

No. 20 Geography Texts for Sixth Forms: A Review of Recent Trends. J. N. Bates. April 1982.

No. 21  Assessment and the Geography teacher. Melvyn Jones. February 1980, revised February 1983.
No. 22  Geography and Environmental Education. Martyn Cribb. May 1983.
No. 23  Geography Computer Software: A Selection of Packages for Use in Secondary Schools. M. J. F. Brown. April 1983.
No. 24  Population geography at A-level. March 1983 (revised edition)
No. 25  Urban Geography. David Bennison and Michael Raw. June 1983.
No. 26  Brazil. John Dickenson and Alison Povall. April 1984.
No. 27  Resources for Mapwork in Primary and Middle Schools. Simon Catling. March 1981 (Revised edition, April 1984).

**Landform Guides** (edited by Rodney Castleden and Chrstopher Green)
(Series produced jointly with the British Geomorphological Research Group.)
No. 1   Classic Coastal Landforms of Dorset. Denys Brunsden and Andrew Goudie. 1981.
No. 2   Classic Landforms of the Sussex Coast. Rodney Castleden. 1982.
No. 3   Classic Glacial Landforms of Snowdonia. Kenneth Addison. 1983.
No. 4   Classic Landforms of the Weald. D. A. Robinson and R. B. G. Williams. 1984.
No. 5   Classic Landforms of the South Devon Coast. Derek Mottershead. 1986.
No. 6   Classic Landforms of the North Devon Coast. Peter Keene. 1986.
No. 7   Classic Landforms of the Gower Coast. E. M. Bridges. 1987.
No. 8   Classic Landforms of the Lake District. John Boardman. 1988.
No. 9   Classic Landforms of the White Peak. Roger Dalton, Howard Fox and Peter Jones. 1988.
No. 10  Classic Landforms of the Northern Dales. Eileen Pounder. 1989.
No. 11  Classic Landforms of the Dark Peak. Roger Dalton, Howard Fox and Peter Jones. 1990.
No. 12  Classic Landforms of the North Norfolk Coast. E. M. Bridges. 1990.
No. 13  Classic Landforms of the Brecon Beacons. R. Shakesby, 1992.

**Fieldwork Location Guides**
No. 1   Brighton and the Downs. C. M. J. Allen. 1981.
No. 2   Maidstone and the Weald of Kent. F. M. Laundon. 1982.
No. 3   Leeds and Bradford. David Skinner and Patrick Wiegand. 1983.
No. 4   The Brecon Beacons. Michael Francis and Nigel Lowson. 1984.
No. 5   Birmingham. John Gerrard and Robert Prosser. 1987.

# References

The main references have been the archival files of the Geographical Association and the Royal Geographical Society together with the Association's Journals—*The Geographical Teacher, Geography, Teaching Geography* and *Primary Geographer*, plus the Association's publications (see Appendix M). Additionally the following references in particular have been consulted:

Bailey, Patrick, *Securing the Place of Geography in the National Curriculum of English and Welsh Schools*, Leicester University Geography Department, Occasional Paper, No. 20, 1991.

Baker, J. N. L., *History of British Geography*, Blackwell, Oxford, 1963.

Balchin, W. G. V., Careers for Graduate Geographers, *The Geographical Journal*, Vol. 149, No. 3, Nov. 1983, pp. 334–341.

—United Kingdom Geographers in the Second World War, *The Geographical Journal*, Vol. 153, No. 2, July 1987, pp. 159–180.

—One Hundred Years of Geography in Cambridge, *Cambridge*, the Magazine of the Cambridge Society, Number 23, 1988, pp. 39–53.

Department of Education and Science (DES), *The Framework of the School Curriculum*, London, HMSO, 1980.

—*The School Curriculum* London, HMSO, 1981.

—*The National Curriculum 5 to 16*, A Consultative Document, London, HMSO, 1987a.

—*A Bill to Amend the Law relating to Education*, London, HMSO, 1987b.

—*National Curriculum Geography Working Group*, Interim Report, London, HMSO, 1989.

—*Geography for ages 5 to 16*. Proposals of the Secretary of State for Education and Science and the Secretary of State for Wales (Final Report of the Geography Working Group), London, HMSO, 1990.

Fleure, H. J., Sixty Years of Geography and Education, *Geography*, Vol. 138, pp. 231–266, November 1953.

—Chairs of Geography in British Universities, *Geography*, Vol. 46, pp. 349–353, November 1961.

Freeman, T. W., *A Hundred Years of Geography 1861–1961*, Duckworth, 1961.

—*A History of Modern British Geography*, Longman, 1980.

Freshfield, D. W., Valedictory Address, *Geographical Teacher*, No. 29, Vol. VI, Pt. 1, Spring, 1911, p. 5.

Garnett, Alice, The Geographical Association in 1964, *Geography* No. 224, Vol. 49, Pt. 3, July 1964, p. 167.

Herbert John Fleure, *Biographical Memoirs of Fellows of the Royal Society*, Vol. 16, November 1970.

Gilbert, E. W. *et al.*, Andrew John Herbertson, 1865–1915. *Geography*, Special Issue No. 229, Vol. 50, Pt. 4, November 1965.

Herbertson, A. J., The Major Natural Regions, *Geographical Journal*, Vol. 25, pp. 300–312, 1905.

Her Majesty's Inspectors of Schools, *A View of the Curriculum, Matters for Discussion No. 1*, London, HMSO, 1980.

—*The Curriculum from 5 to 16, Matters for Discussion No. 2*, London, HMSO, 1985.

—*Geography from 5 to 16, Curriculum Matters No. 7*, London, HMSO, 1987.

Keltie, J. Scott, Thirty Years Progress in Geographical Education, *The Geographical Teacher*, No. 38, Vol. 7, Pt. 4, Spring 1914, p. 215.

Mackinder, H. J., *Britain and the British Seas*, 2nd Edition, Oxford, 1902.

Mill, H. R., *The Record of the Royal Geographical Society*, The Royal Geographical Society, London, 1930.

National Curriculum Council (NCC), *Geography 5 to 16 in the National Curriculum*. A Report to the Secretary of State for Education and Science in the Statutory Consultation for Attainment, Targets and Programmes of Study in Geography, York, NCC, 1990.

Scargill, D. I., The R.G.S. and the Foundation of Geography at Oxford, *Geographical Journal*, Vol. 142, Pt. 3, November 1976, p. 438.

Steel, R. W., *The Institute of British Geographers—The First Fifty Years*, IBG, London, 1984.

Stoddard, D. R., The R.G.S. and the Foundation of Geography at Cambridge, *Geographical Journal*, Vol. 141, Pt. 2, July 1975, p. 216.

Taylor, E. G. R., *The Mathematical Practitioners of Tudor and Stuart England*, 1930.

Wise, M. J., The Scott Keltie Report 1885 and the Teaching of Geography in Great Britain, *Geographical Journal*, Vol. 152, Pt. 3, November 1986, p. 367.